National Park Service
U.S. Department of the Interior

Fort Stanwix National Monument
Rome, New York

Fort Stanwix National Monument: *Alternative Transportation Study*

PMIS No. 106753
November 2010

John A. Volpe National Transportation Systems Center

U.S. Department of Transportation
Research and Innovative Technology Administration

Table of Contents

Acknowledgments

Mathew J. Andrews, Planning Coordinator, city of Rome, Department of Community Development

Debbie Conway, Superintendent, Fort Stanwix National Monument

Diane Shoemaker, Director of Community Development, city of Rome, Department of Community Development

Executive summary

The Fort Stanwix National Monument Alternative Transportation Feasibility Study is a publication of the U.S. Department of Transportation's (DOT) John A. Volpe National Transportation Systems Center in Cambridge, Massachusetts. Sponsored by the National Park Service's (NPS) Northeast Region Office, the project was funded with a grant from the Paul S. Sarbanes Transit in Parks Program.

Located in the center of downtown Rome, New York, Fort Stanwix currently experiences several transportation-related issues affecting visitor access to the park and overall visitor experience. As a follow-up to a 2006 Alternative Transportation Study, the goals of this report are threefold: 1) to update the 2006 study's existing conditions report to reflect recent changes that have occurred in and around the park, 2) to evaluate five specific areas of concern to the park, which were identified in the 2006 study, and 3) to identify opportunities to a) address these areas of concern; b) improve visitor experience at the park; and c) strengthen the park's relationship with the city by furthering city and county goals. The five areas of concern to the park are nonmotorized trail connections, vehicular signage and wayfinding, parking, pedestrian access, and shuttle feasibility. Findings for each of these focus areas are summarized below.

Nonmotorized trail connections

Fort Stanwix hopes to attract visitors traveling along local and regional trail networks, provide recreational opportunities for visitors, and increase public awareness of and access to nearby historic sites and attractions. This section of the report focuses on the connection of the existing New York State Canalway Trail in Rome, the alignment of the planned Mohawk Trail along the Mohawk River, and several additional local nonmotorized transportation opportunities in and around Rome.

This section emphasizes cooperation among trail stakeholders and supports the concept of Fort Stanwix serving as an important destination and connection point for nonmotorized travelers. The study team generally recommends off-road dedicated rights-of-way for trails as a long-term goal. In the near term, wide sidewalks for pedestrians and on-road painted bike lanes will allow safer passage through Rome's city streets.

Signage and wayfinding

New signage along major interstates, regional corridors, and local roads will improve the visitor experience of traveling to the park while ensuring that first-time visitors and tour bus drivers are able to more easily navigate to the park and its parking facilities. Currently, signs are not present in key locations surrounding the city and are difficult to see in the immediate vicinity of the park. Improved vehicular wayfinding will reduce the number of confused out-of-town drivers in downtown Rome, making the entire area safer for those choosing to travel on foot or bicycle.

This section of the report focuses on existing sign locations and identifies opportunities for new vehicular directional signs within a 15-mile radius of the park. This section identifies and prioritizes 17 new signs that would benefit motorists traveling to Fort Stanwix. The section also includes summaries of the steps necessary to develop new signs, partners needed to install the signs, and approximate cost of new signs.

Parking

Issues related to parking at Fort Stanwix stem from the fact that all parking nearby options around the park are owned and managed by entities other than NPS. As a result, the park has little control over parking location, fees, accessibility, signage, facilities, management, and maintenance. Furthermore, the park is unable to assure visitors that their cars will not be ticketed or towed, that their visits will not be cut short due to time restrictions, or that their vehicles are secure. Finally, a permanent parking arrangement would allow Fort Stanwix to focus improvement efforts on pedestrian safety between the parking area and the visitor center.

This section of the report identifies current parking arrangements and practices, assesses future parking needs, and identifies, evaluates, and recommends parking alternatives. The study team recommends that

Fort Stanwix should work with the city to investigate opportunities or agreements that could ensure that the city-owned Fort Stanwix parking garage is available for parking seven days a week, year-round. Though it may require a significant effort in order to change the city's current garage policies, the result could produce an impact that positively affects all parties.

Pedestrian access and safety

This section of the report highlights potential improvements to pedestrian-oriented infrastructure and provides a list of prioritized recommendations. Special attention is given to major intersections, entrances, and along sidewalks and pathways in and around the park.

The most important recommendation in this section involves working with local stakeholders to create a unified pedestrian improvement zone in conjunction with the city's economic development corridors. The study team also emphasizes additional safety improvements to the intersection of North James Street and West Dominick Street, with a recommendation of traffic controls to stop vehicle traffic in order to achieve maximum pedestrian safety in one of the city's most walkable areas.

Shuttle opportunities

The shuttle section explores the possibilities of land-based shuttles, water-based shuttles, and amphibious shuttles between Fort Stanwix and Oriskany Battlefield State Historic Site. After an examination of current transportation trends and infrastructure and an acknowledgment that visitation numbers at Oriskany are not available, the report does not recommend long-term investment in a shuttle program. Should the park reach a point in time when demand for shuttle service can be proven, the study team recommends a short-term pilot, during the peak visitation season, with minimal investment required by NPS or other stakeholders.

Existing transportation conditions update

Overview

Project overview

This section represents an update to the 2006 Volpe National Transportation Systems Center Fort Stanwix National Monument transportation summary report. The purpose of this section is to inventory and identify existing conditions to inform and serve as a foundation for the evaluation of the five focus areas of this report: nonmotorized trail connections, signage and wayfinding, parking, pedestrian access and safety, and shuttle opportunities.

The 2006 summary report provides an overview of transportation conditions, regional attractions, transportation networks, and future transportation recommendations for Fort Stanwix National Monument. The report summarizes visitation statistics, local attractions, pedestrian and vehicular access, and transportation infrastructure.

This existing conditions update reviews transportation changes within the region and the park since the summary report was issued in 2006. Since 2006, there are new National Park Service (NPS) and city of Rome reports, plans, trends, and site changes that are transforming transportation and park access. The most notable report is the 2009 Fort Stanwix National Monument Final General Management Plan (GMP) and Environmental Impact Statement (EIS), which outlines the park philosophy for planning for the next fifteen to twenty years. Another important change since 2006 is that the park now co-manages (with the state of New York) Oriskany Battlefield State Historic Site, which lies approximately six miles to the southeast by car. Similar to the summary report, this update report is organized by topics, which are divided into current and future conditions.

Fort Stanwix and the city of Rome

Fort Stanwix is in downtown Rome, New York, (population 33,805) within Oneida County (population 234,649[1]) (Figure 1). The population of Rome and Oneida County has been decreasing since the 1990 U.S. Census, when the population of the city was 44,350 and the population of the county was 250,836. The decrease in population is in part due to the closure of Griffiss Air Force Base in 1990. In order to reinvigorate the city's downtown area, the city of Rome is investing in parks, downtown streetscapes, and regional connections. Since Fort Stanwix is in downtown Rome and adjacent to the city's center, many of the city's investments will improve the overall visitor experience to the park.

---

[1] U.S. Census Bureau, 2008-2010 American Community Survey

Figure 1
Rome context map
Source: Volpe Center, New York State Department of Transportation (NYSDOT)

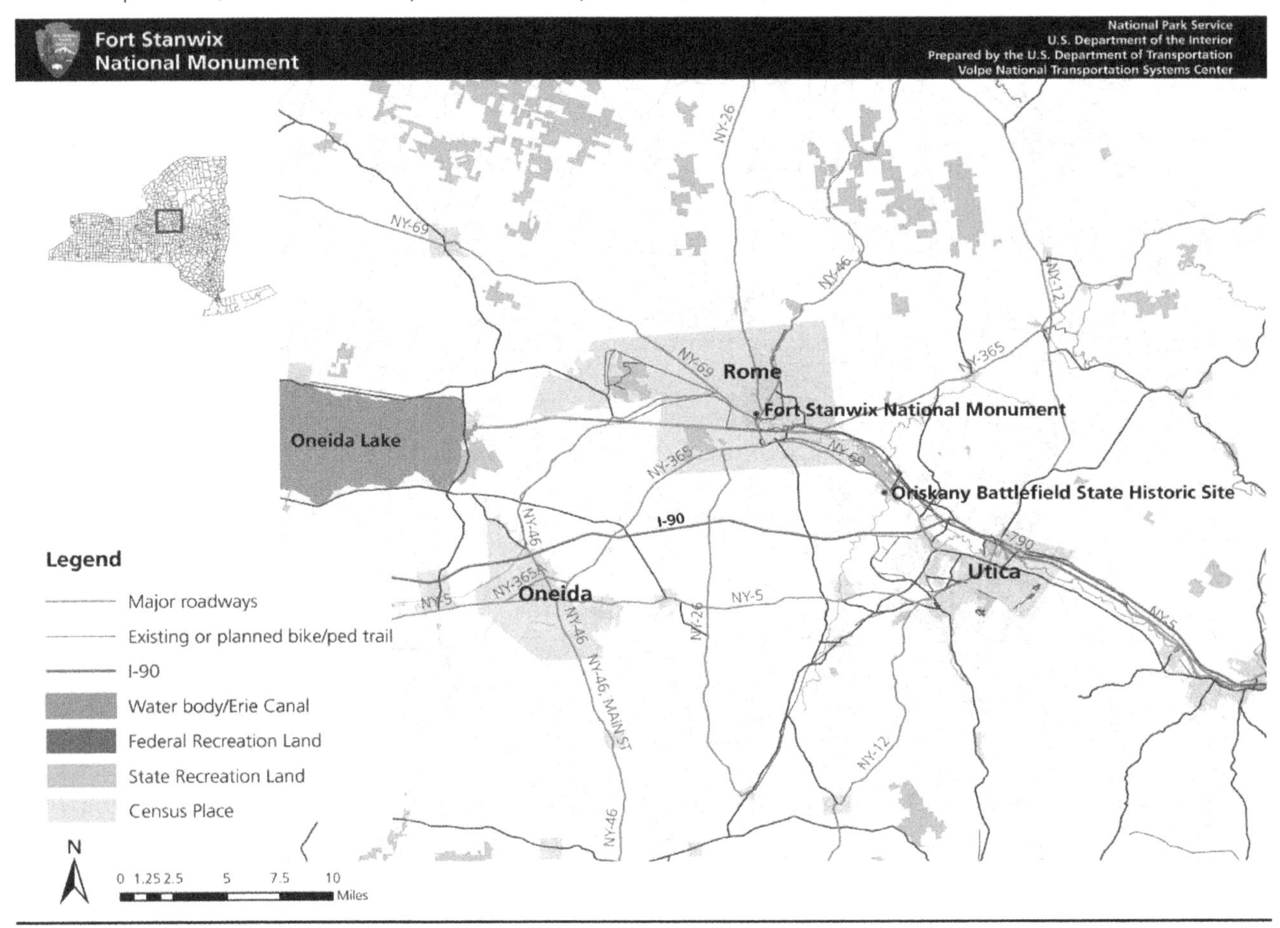

The boundary of the park remains the same as the 2006 park boundary. Much of the surrounding land use, street layout, and approach to the site remains the same (Figure 2). Transportation continues to play an important role at the park because of the multiple modes of transportation that provide access to the park and downtown Rome. Vehicular traffic on highways and roads remains the most common travel mode in the area; however, canals, rail lines, public transportation, pedestrian and bicyclist trails, and city streets and sidewalks all offer transportation options for visitors traveling to downtown and the park. The desire to attract more visitors to the region continues through local efforts by the park, the city of Rome, Oneida and Herkimer counties and their metropolitan planning organization, and other local groups.

Figure 2
Rome and Fort Stanwix
Source: Volpe Center, NYSDOT

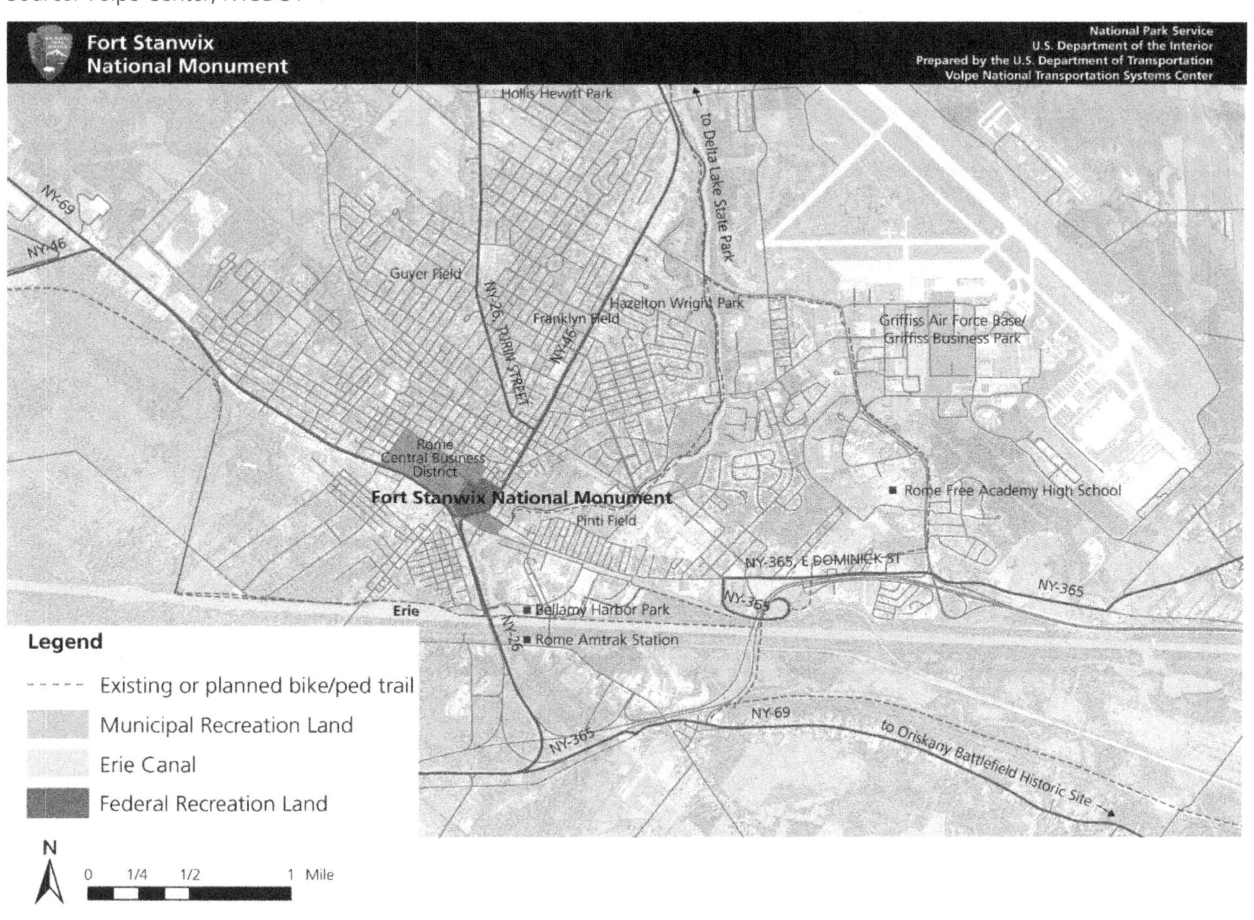

Fort Stanwix National Monument's sixteen acres include the park, which is a reconstruction completed in 1976, the Willett Visitor Center, an additional interpretive center within the park itself (which used to be the main visitor center), a maintenance facility, and surrounding grounds. Following the completion of the Willett Center in 2005, the park continued to make site improvements, including new walkways, natural planting areas, and outdoor educational space at the entrance to the visitor center. Visitors to the Willett Center have the opportunity to watch a series of short films recounting the story of the Siege of Fort Stanwix as seen through the eyes of four historical personalities, and can view artifacts that were recovered at the site during the park's reconstruction.

Visitation

Fort Stanwix welcomed 103,748 visitors in 2010, an increase of more than 30,000 visitors from 2008 (Figure 3) and an increase of more than 40,000 visitors from 2007.[1] Fort Stanwix and the Willett Center are open year round except for Christmas and New Year's Day. The Willett Center is open year round from 9:00 AM to 5:00 PM. In December, January, February, and March, the park is open for three guided tours a day at 10:30 AM, 12:30 PM, and 2:30 PM. From April through November, the park is open from

[1] National Park Service Public Use Statistics Office. Annual Park Visitation. http://www.nature.nps.gov/stats/park.cfm?parkid=493. Visitors are counted as they enter Fort Stanwix.

9:00 AM to 4:45 PM. Park visitation is highest in July and other popular months include June, August, and September (Figure 4).

Figure 3
Annual visitation
Source: NPS and Volpe Center

Figure 4
Visitation by month
Source: NPS and Volpe Center

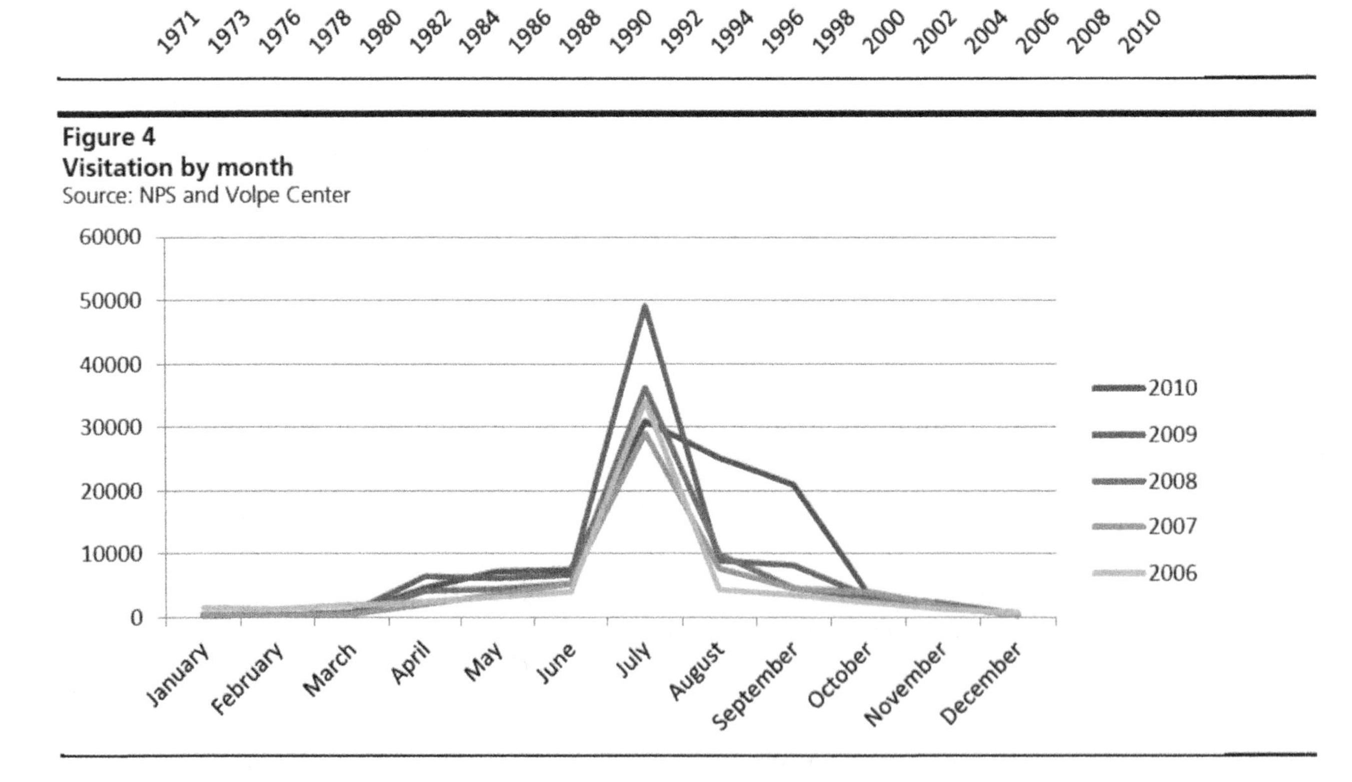

In addition to daily tours, the park holds special events that attract visitors interested in specialized programming related to military reenactments, historical weaponry, Native American culture, and general ways-of-life during the colonial era. During summer months, the park frequently hosts weapon demonstrations as well as family-oriented educational events that are free to the public. On the Fourth of July, the park hosts public readings and signings of the Declaration of Independence. On August 6, the park hosts an evening ceremony and wreath-laying to commemorate the Battle of Oriskany. In late

August, the park hosts Founder's Day, which includes programs and junior ranger activities throughout the day. In October, the park conducts a candlelit tour that includes wilderness stories of life in the park. "Wintering at the Fort" is a program allowing visitors to learn about life inside the park during the winter months, and the park hosts candlelight tours, including caroling and children's events, during the holiday season.

General Management Plan

In 2009, Fort Stanwix National Monument completed a General Management Plan (GMP). The two management alternatives in the GMP remain similar to information that was initially reported in a 2006 summary report from the GMP. Fort Stanwix managers hope to clarify the park's boundary, which is currently bounded by North James Street, Black River Boulevard, Erie Boulevard, and Park Street. The park plans to promote the cultural and historical significance of Fort Stanwix by highlighting the connection to the canal and the surrounding historical sites. The park aims to design pedestrian routes within and around the park so that it is more accessible to those traveling on foot. Improving access is also stressed within the park itself. The park hopes to invest in existing infrastructure and facilities to make existing buildings more efficient. Last, the park aims to improve wayfinding and signage for visitors.

As stated in the 2006 summary report, Alternative 1 of the GMP is the no action alternative for Fort Stanwix. The no action alternative serves as a reference for all proposed changes. With this alternative, the park plans to update interpretation and programming and to strengthen partnerships. Under the no action alternative, there is no possibility for new construction within the park; however, the park will work to link downtown Rome, surrounding trails, the Erie Canal to the park.

Alternative 2 of the GMP was approved by the Regional Director of the Northeast Region on December 16, 2009. Alternative 2, the park's preferred action alternative, expands interpretation within the region, restores landscape areas, permits the reconstruction of historically significant structures, and utilizes existing park space for additional public use. In this preferred action alternative, Fort Stanwix will work closely with other historically significant sites to create a more comprehensive visitor experience. The preferred action alternative recommends the use of traffic controls at street crossings, physical links to nearby public trails, and shuttle connections between historic sites. The park will require additional staff to provide visitor services and programming. Fort Stanwix will work to improve partnerships with other sites in the region, and will partner with new agencies to coordinate programming and community outreach.

Area attractions

Many of the area attractions described in the 2006 Summary Report remain the same. However, the Rome Community and Recreation Center, mentioned in the 2006 Summary Report, is not going to be constructed in the near future.

Bellamy Harbor Park

Situated on the banks of the Erie Canal less than a mile south of downtown Rome, Bellamy Harbor Park (Figure 5) is a valuable recreational amenity for local residents while serving as a unique gateway to the Rome area for visitors traveling on the Erie Canal. Owned and managed by the New York State Canal Corporation and a component of the Erie Canalway National Heritage Corridor, the park features a boat dock, concrete promenade, pedestrian paths, and a pedestrian bridge over the Mohawk River. A new information and wayfinding center is planned for the park, with the goal of helping out-of-town visitors access destinations and shops in town. In addition, a new pavilion will include updated facilities and a renovation of the existing boat docks. Plans for the pavilion were developed with funds from the New York State Department of State and the New York State Canal Corporation.

Oriskany Battlefield State Historic Site

Oriskany Battlefield State Historic Site (Figure 5), located a little over five miles southeast of Fort Stanwix, is currently co-managed by the National Park Service and the New York State Office of Parks, Recreation, and Historic Preservation. In 2008, NPS entered into an agreement to provide staffing and management of the site during summer months, while both organizations contribute to maintenance. The Battle of Oriskany played an important role in the siege of Fort Stanwix, and the new agreement allows NPS to provide programming and services that better connect the two sites. Currently there is no signed vehicle, bicycle, or pedestrian route that connects the two sites.

Figure 5
Bellamy Harbor Park (left) and Oriskany Battlefield State Historic Site (right)
Source: Volpe Center

Steuben Memorial State Historic Site

In the same agreement whereby NPS assumed day-to-day management of Oriskany Battlefield, NPS also agreed to provide assistance in managing the Baron von Steuben Memorial State Historic Site. Situated approximately eighteen miles northeast of Rome, the memorial honors Baron von Steuben for his Revolutionary War contributions. The site includes a re-constructed log cabin, a sacred grove, and a large monument marking Steuben's grave.

Heritage corridors and scenic byways

Erie Canalway National Heritage Corridor's 2006 Preservation and Management Plan details the Heritage Corridor's strategies for managing the Erie Canal corridor, including both water- and land-based trails (Figure 6). The plan includes several management goals, including the protection of the corridor's historic sense of place and natural resources, provision of recreation opportunities, promotion of economic growth and heritage development, and the attraction of American and international visitors.

Figure 6
Erie Canalway National Heritage Corridor, western portion (top), eastern portion (bottom)
Source: NPS

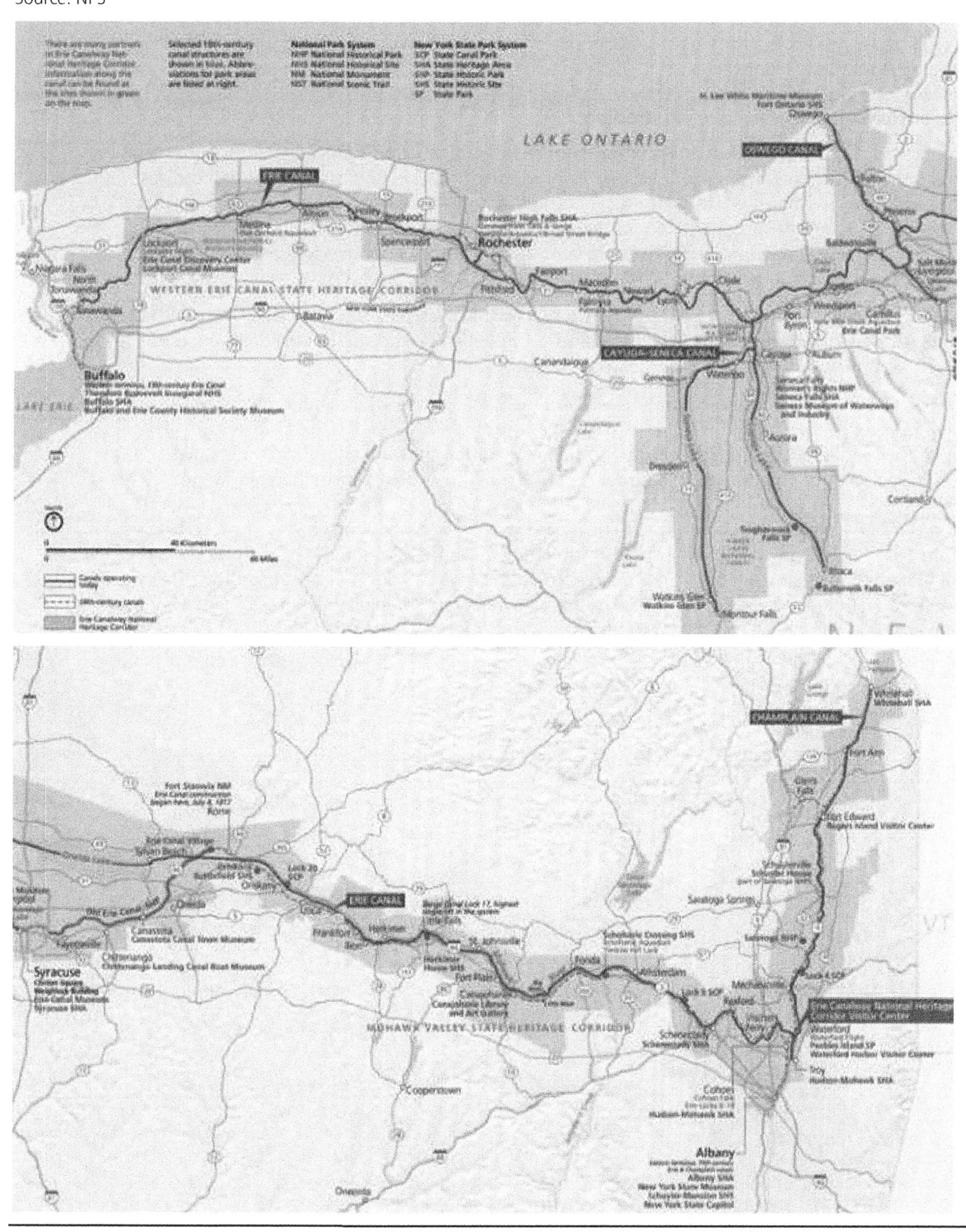

Three New York State Scenic Byways are connected to the Rome area. These byways include the Black River Trail that connects Rome to Ogdenburg, the Central Adirondack Trail that connects Rome to Glen Falls, and the Revolutionary Trail that connects Albany to Lake Ontario. The Scenic Byways Program produces road maps listing scenic attractions and lodging recommendations for drivers using the scenic routes. A sign for the start of the Central Adirondack Trail is located on the Fort Stanwix boundary along Black River Boulevard (Figure 8).

Traffic, circulation, and vehicular wayfinding

The roads and traffic around Fort Stanwix have not changed significantly since the 2006 transportation summary report. The convergence of Black River Boulevard and Erie Boulevard, known locally as Spaghetti Junction, remains confusing for drivers and difficult for pedestrians and bicyclists. The 2006 transportation summary report included preliminary urban design plans for the Spaghetti Junction intersection; however, planning and construction of the new intersection has not begun and is not expected to commence in the near term.

Traffic levels throughout the greater Utica-Rome area are illustrated in Figure 7 using the most recent available AADT (annual average daily traffic) counts. Immediately adjacent to Fort Stanwix, traffic levels have increased slightly since 2000, as indicated in Table 1.

The park's website provides driving directions from the west, east, northeast, and northwest, also shown as a map in Figure 7. In written form, these directions appear complicated, in part due to the physical nature of the road network, but also because of the many different route numbers, road names, abbreviations, and punctuation used. A direct excerpt from the website is provided below:

> From the East:
>
> Take I-90 (NY Thruway) to Exit 32 (Westmoreland). Turn Right onto Rt. 233 N. Follow Rt. 233 for 6 miles, and then make a left onto the Rt. 49N/Rt. 69W/Rt. 365W ramp. Continue on Rt. 49N for 0.1 mi. and bear right on ramp to Rt. 49N/Rt. 69W first 0.2 miles, and then 1.2 miles. Turn right onto Rt. 26 (N James St.). The park will be on your right.

Figure 7
Rome region driving routes to the park and AADT
Source: NYSDOT and Volpe Center

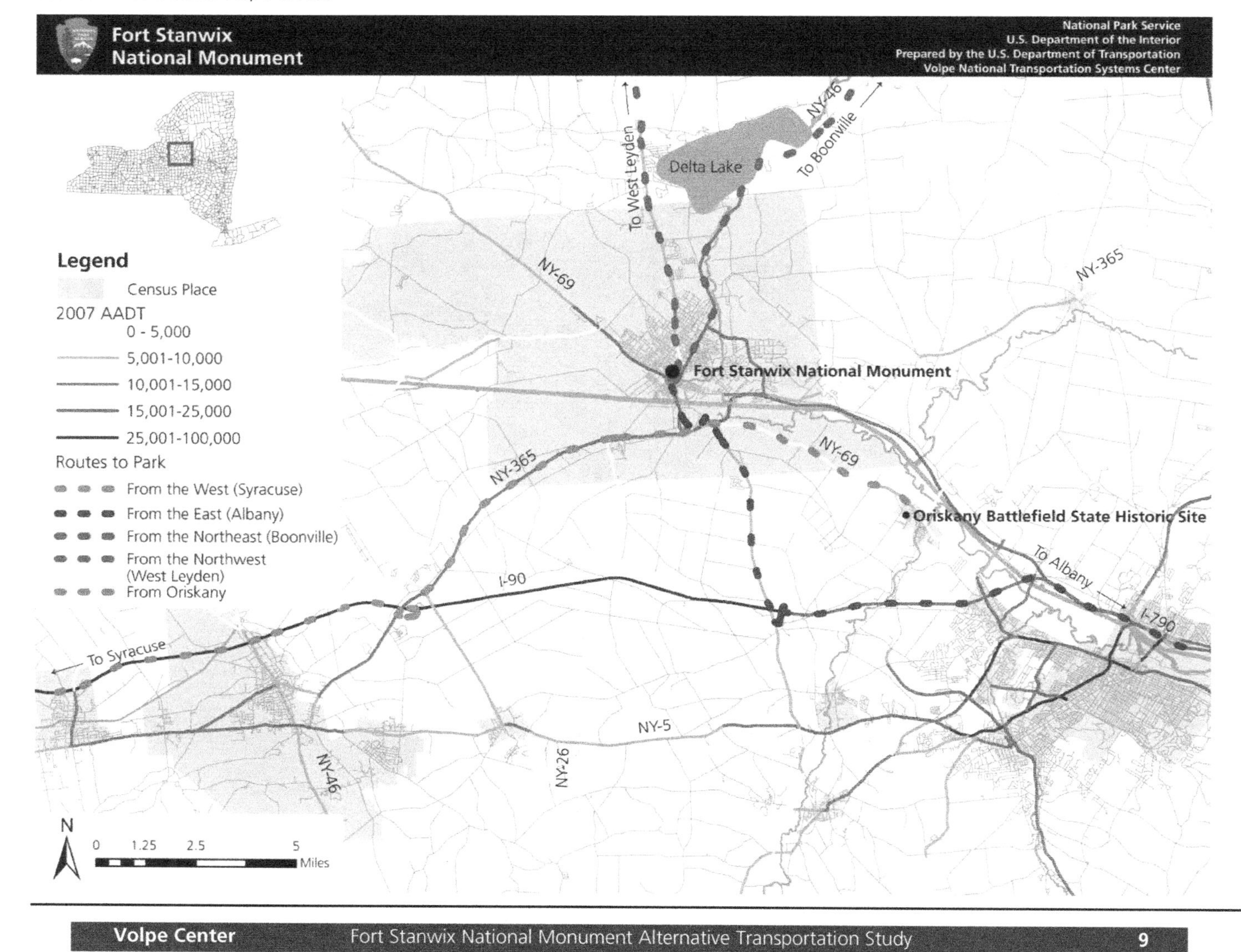

Table 1
Rome region AADT
Source: NYSDOT and Volpe Center

Roadway	AADT 2000	AADT 2008
Erie Boulevard	21,500	22,129
Black River Boulevard	21,000	22,505
North James Street	8,500	9,226

Few signs exist to guide drivers between Interstate 90 and downtown Rome. In particular, signage to Fort Stanwix from the interstate is inconsistent and does not always provide a clear direction to visitors. Once a vehicle enters downtown Rome, several wayfinding signs are situated along James Street and Black River Boulevard (Figure 8), but directions to parking or pull out areas are few. Existing signs are small and often clustered with several attractions on one pole. As stated in the 2006 transportation summary report, signage at the infamous Spaghetti Junction attempts to communicate too much information.

Despite difficulties that a visitor might have finding his or her way to Fort Stanwix or downtown Rome, NPS signage for the park itself is easily visible when passing by the site on any of the major travel corridors (Figure 9).

Figure 8
Examples of vehicle wayfinding signs: James Street (left), Black River Boulevard (center), Martin Street (right)
Source: Volpe Center

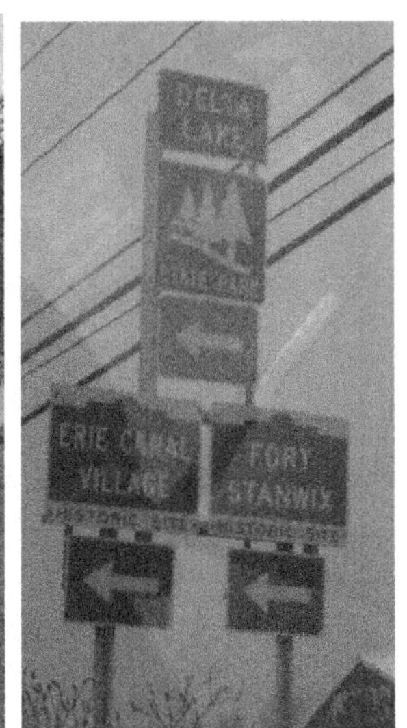

Figure 9
Fort Stanwix signs on James Street (left) and on Black River Boulevard (right)
Source: Volpe Center

Parking

Fort Stanwix does not own a public parking facility, nor does it have designated spaces for visitors in nearby lots. The NPS directs visitors to the Fort Stanwix Parking Garage (Figure 10), which is owned and operated by the city. Parking in the garage is convenient to access the visitor center; however, the parking garage is not open most weekends and, according to the city, is often full during the week. Furthermore, the cost of parking in the garage is ten dollars per day. In the summer of 2009, the parking garage was open on weekends, but it remains to be determined if this service will continue in the future.

Two bus parking spaces are available along North James Street, and the park tries to manage the arrival of large groups so that this parking is available. However, these spaces are sometimes full, so buses or RVs must find alternative spaces.

In addition to the parking garage, several parking lots near Fort Stanwix are not owned by the NPS. Many visitors park in the Rome Savings Bank parking lot, at West Dominick Street and North James Street, because it is directly across the street from the Willett Center. This lot is uncovered and able to accommodate RVs and buses. However, in winter, snow is not removed, and visitors are unable to access the parking lot. In previous years, NPS had worked out an informal agreement with the bank to plow the lot in the winter so that the bank and the park could both use the lot. Recently, the bank decided to discontinue this arrangement, primarily due to liability concerns. If the parking lot is put up for sale, the park is interested in the possibility of buying or creating a more formal agreement to plow the lot in exchange for wintertime use.

Several on-street parking spaces are located north of the park, adjacent to the park administration and Rome Historical Society building. Parking is limited to ninety minutes, although the city does not typically ticket cars that park in this area for longer than that amount of time. There is also street parking along West Dominick Street, but the two-hour time limit along that street compromises the amount of time a visitor can spend visiting the park.

Another parking lot on Black River Boulevard serves a laundromat, salon, and pizza shop. While this lot is close to the Visitor Center, there is no safe place to cross the street. South of the park and along Black River Boulevard, there are several private parking lots; however, there is no safe pedestrian access to the park. Finally, there are numerous small private parking lots between Erie Boulevard and West Dominick Street. The lots vary in size, occupancy rate, access, and connectivity to the park and other destinations. None of these lots provides the same or better access than that of the parking garage or bank parking lot.

Figure 10
Fort Stanwix Parking Garage
Source: Volpe Center

Bus service

Public transportation in Rome is operated by Centro of Oneida, part of the Central New York Regional Transportation Authority (CNYRTA). Centro operates on a fixed route and includes demand response service within the city. Six of Centro's Rome routes provide access to Fort Stanwix, the Amtrak train station, Griffiss Technology Park, shopping, and the hospital (Figure 11). Each route travels within a block of the park, there is a bus stop by the Rome Savings Bank for four routes, and a major transfer site at 225 Liberty Street is three blocks northwest of the park. The bus runs mid-week with a limited schedule on Saturdays; it does not operate on Sundays and select holidays. In addition to the main routes, paratransit service is provided to disabled passengers.

Figure 11
Rome bus routes
Source: Central New York Regional Transportation Authority, 2009 http://www.centro.org/

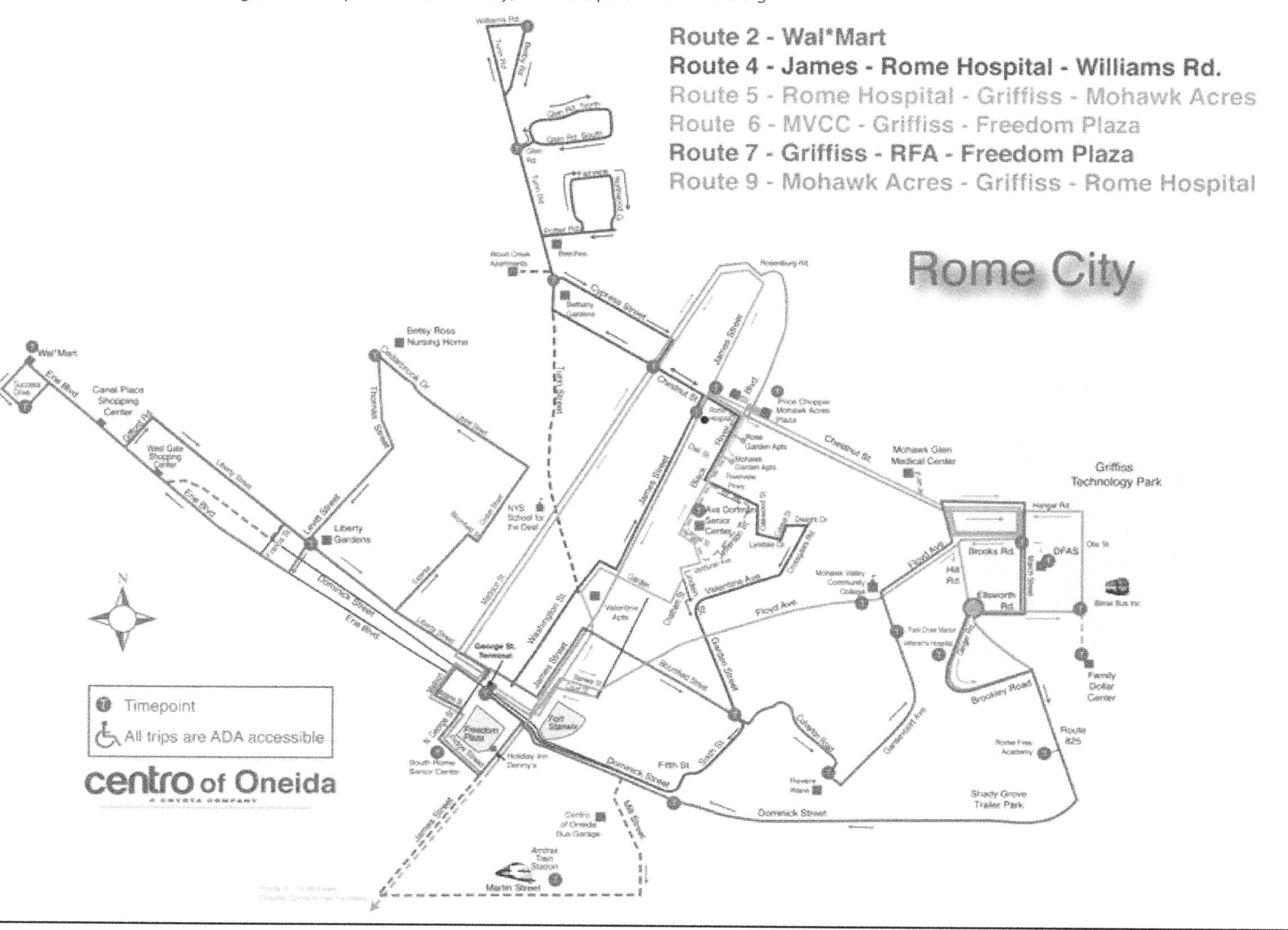

In Oneida County, Centro reports that ridership was up in 2008 from 2007 due to a rise in gas prices and improved service.[1] In Rome, ridership increased from 39,688 in 2007 to 41,113 in 2008. In addition, ridership increased in Utica from 245,466 in 2007 to 278,786 in 2008. The fare for a regular ride is $0.75 and a monthly pass is $25.50. Student rates are $0.50 and a monthly pass is $17.00. A ride for a senior citizen costs $0.35 and a monthly pass is $12.00.

The Birnie Bus Service provides interurban transportation in central New York. In addition, Birnie Bus provides services within Oneida County and operates Monday through Friday.

Streetscape improvements

As indicated in the city of Rome's Comprehensive Plan (2003) and Urban Design Plan (2006), the city's streetscape improvement efforts focus on three corridors: West Dominick Street, East Dominick Street, and North James Street.[2] All three corridors converge in the center of town at Fort Stanwix.

West Dominick Street

As the heart of Rome's central business district, West Dominick Street is home to many businesses and civic institutions, including City Hall. In 2003, Rome's Comprehensive Plan described downtown as "characterized by poor visibility, confusing traffic patterns, limited pedestrian access, considerable underutilized space and no strong visual 'anchor'." Conditions do not appear to have dramatically improved, and vacant storefronts are common. The eastern terminus of the West Dominick Street corridor is at the intersection with North James Street, which is also the western boundary of Fort Stanwix.

North James Street

According to the city, businesses along North James Street have taken advantage of the city's façade improvement program, wherein small businesses are able to apply for low-interest loans to improve storefront conditions and general curb appeal. Today, a number of businesses and restaurants operate in a small business district located a few blocks north of downtown.

East Dominick Street

Newly designated Rome's Little Italy, the East Dominick Street corridor runs from Black River Boulevard (Fort Stanwix's eastern boundary) to Nock Street, transitioning from a small, traditional commercial district to a unique mix of businesses and residences. Unlike West Dominick, East Dominick Street retained a number of historic structures with active restaurants and night spots housed inside. However, vacancies also exist along this corridor. Traveling east in this corridor, the charm of Little Italy transitions to a more industrial area, but streetscape improvements and themed signage continue beyond the pedestrian-friendly part of this district.

Pedestrian and bicycle facilities

Urban streets

Given Fort Stanwix's location in the center of Rome, sidewalks and pathways can be found both in and around the park, and entrances are well marked with recognizable NPS signage (Figure 9). Fort Stanwix's central location affords fairly safe and direct access from each of the city-designated development corridors (described above in Streetscape Improvements). Pedestrian access from the south is more difficult, however, due to the complex intersection of Erie and Black River Boulevards at Spaghetti Junction.

---

[1] Centro. "What's New At Centro?" http://www.centro.org/whats-new.aspx
[2] City of Rome Comprehensive Plan, 2003.

A new painted intersection at North James Street and West Dominick Street is a clear crossing location for pedestrians and visitors to the park (Figure 12). Visitors parking in the Fort Stanwix Parking Garage or the Rome Savings Bank parking lot are able to cross North James Street at this point to access the Willett Visitor Center. The new painted intersection consists of textured and colored pavement in the shape of the star fort as well as pedestrian crossing caution signs in the middle of the road.

The city owns the sidewalks along North James Street that connect the park entrance to downtown Rome. Along this corridor, the city recently installed new lighting and light post banners helping to identify the neighborhood. Within the park boundary, wide gravel pathways connect the visitor center to the park (Figure 12).

Figure 12
Fort Stanwix Pedestrian Facilities: James Street crossing (left), James Street (center), and park paths (right)
Source: Volpe Center

Multi-use trails

A number of multi-use trails have been developed within Rome and in the surrounding region (Figure 13). A component of the Erie Canal National Heritage Corridor, the Erie Canalway Trail is a series of completed and planned trail segments running the length of the historic Erie Canal from Albany to Buffalo. The existing paved and crushed stone trail from Rome southeast to Utica (Figure 14) is fourteen miles. Walking, bicycling, and cross country skiing are the most common activities along the trail. The path from Rome southwest to Dewitt is within the Old Erie Canal State Park Trail, and the distance is thirty-six miles. The path is crushed stone, and horseback riding, snowmobiling, walking, bicycling, and cross country skiing are the most common activities. While the trail has been completed from both the east and west sides of town, the trail was never completed in central Rome. Rather, signs have been installed to direct trail users through Rome along city streets.

Figure 13
Herkimer Oneida County multi-use trails
Source: Herkimer Oneida County (HOC), Volpe Center, NYSDOT

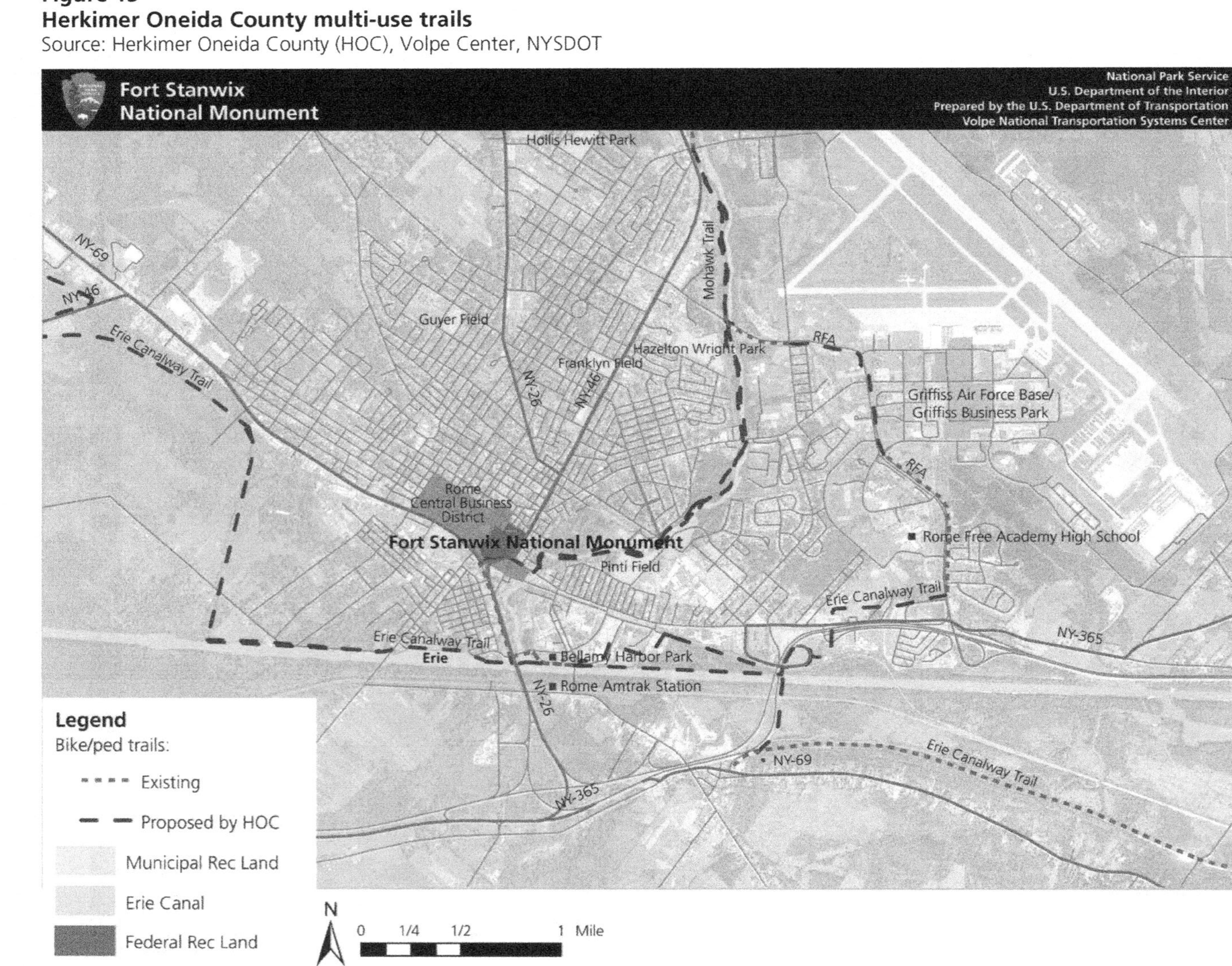

Figure 14
New York State Erie Canalway Trail access points near Oriskany Battlefield State Historic Site
Source: Volpe Center

The Mohawk River Trail is a proposed multi-use trail that will connect Bellamy Harbor Park with Delta Lake State Park north of Rome. The city of Rome is investigating potential connections between Bellamy Harbor Park and Fort Stanwix. One possibility is to use a swath of undeveloped National Grid land along the Mohawk River. In this scenario, Fort Stanwix would serve as an intermediate destination, providing direct bicycle/pedestrian connection to the park from Bellamy Harbor Park and Delta Lake State Park. Alternative alignment options that do not provide access to Fort Stanwix have also been considered.

Further proposed trails and existing trail segments have been illustrated in the Herkimer Oneida County Bike Atlas, published in 2007 (Figure 13), though these alignments have not been agreed to by all stakeholders.

Currently, signage is not oriented toward pedestrians and bicyclists arriving to the park from the Erie Canalway Trail or from downtown Rome. In Bellamy Harbor, an unreadable posted map needs to be replaced in order to clarify the park's vicinity to the harbor. Once pedestrians and bicyclists are near the park, the approach to the grounds of Fort Stanwix is clear from James Street and from Black River Boulevard. Two large NPS signs adjacent to the sidewalks (Figure 9) direct visitors to the visitor center and park, and two bicycle racks – one just outside of the Willett Center and the other just outside the park itself – are available to visitors. A bicycle rack is also available at Oriskany Battlefield.

Erie Canal

One of Rome's most unique amenities is the Erie Canal. First completed in 1825, the canal is now used primarily by recreational boaters. The 524-mile New York State Canalway Water Trail is an initiative to establish a coordinated "trail" with boat launches and campsites along the New York State Canal System, encouraging more types of motorized and non-motorized boaters to explore this historic resource. Though limited freight transport still occurs on the canal, demand for outdoor recreational access to the canal has increased in the past decade.[1]

Rail service

The Rome Amtrak station is about a mile south of Fort Stanwix on Martin Street. There is one Centro bus route, with service every half-hour to one hour during the week, that connects to the Amtrak station. The

[1] New York State Canal Corporation. http://www.ntscanals.gov

station is on the Empire Service, which connects New York City to Niagara Falls, and the Maple Leaf line, which connects Toronto to New York City. The Maple Leaf line offers a Trails and Rails program that operates from the Albany-Rensselaer station west to Syracuse and is based on the Erie Canalway National Heritage Corridor. The program covers the history of the Mohawk Valley, including the story of the Erie Canal. The program is made possible with additional guidance from the Mohawk Valley Heritage Corridor, the New York State Canal Corporation, and NYSDOT. One-way fares from Albany to Rome are $23.00 in the morning and $33.00 in the afternoon. One-way fares from Penn Station to Rome are $54.00. Three passenger trains depart daily from Rome to New York City.

In January 2010, President Barack Obama's administration announced that upstate New York will receive $148 million in federal stimulus grants to develop a high-speed rail line that will run across the state. The funding will cover a series of track, grade-crossing, and station improvements that will pave the way for a long-term project to bring high-speed rail service to the 285-mile Empire Rail Corridor, which stretches from Buffalo to Albany. This project, which will take approximately 20 years to complete, is expected to increase train speeds to 110 mph on new tracks to be built along existing right-of-way.[1]

Aviation

Griffiss International Airport, owned by the Oneida County, is a user-fee airport with fixed-base operations provided by a private company. The airport is located at the former Griffiss Air Force Base, which closed in 1995.

Conclusion

Fort Stanwix is at the intersection of several driving, bicycling, and pedestrian corridors, and has the opportunity to attract additional visitors already traveling along these routes. The park's location within the city of Rome can benefit the city by attracting visitors to spend time in the city after a visit to the park. This additional visitation can provide economic improvement to the city's declining population. Without sufficient trail infrastructure, connectivity, and signage, the park will have a difficult time attracting new visitor groups. In addition, without a safe arrival and designated parking, visitors will be less likely to stop at the park or remain in the city for an extended amount of time. The transportation infrastructure should provide a safe and clear approach to the park for all user groups with connections to surrounding amenities.

Document review

Below is a list of documents reviewed as part of the update report.

Herkimer-Oneida Counties, *Destinations 2030: Herkimer-Oneida Counties Long Range Transportation Plan* (2009) – Herkimer and Oneida Counties completed this plan to coordinate transportation planning within the two counties. The Herkimer-Oneida Counties Governmental Policy and Liaison Committee is the region's metropolitan planning organization. The 2030 Long Range Transportation Plan promotes an intermodal transportation system with an emphasis on safety and efficiency. The plan lists goals for transportation, including bicycle, pedestrian, rail, and aviation objectives.

Herkimer-Oneida Counties Transportation Study's *Herkimer and Oneida Counties (2007) Bicycling Atlas* (2007)[2] – This atlas includes the locations of bicycle routes throughout Herkimer and Oneida Counties as well as information on state bicycle laws, tips, resources, and safety recommendations. There are numerous state routes, the Erie Canalway Trail, and other multi-use trails available within the counties for bicyclists.

---

[1] Weiner, Mark. "President Obama sends $148 million to Upstate New York for high-speed rail." The Post-Standard. January 28, 2010.

[2] Herkimer-Oneida Counties Transportation Study conducts and performs transportation planning in those two counties

Herkimer-Oneida Counties Transportation Study's *Coordinated Human Services Transportation Plan for Herkimer and Oneida Counties (2008)* – This plan report includes plans, coordination efforts, human services coordination, transportation services, transportation gaps, and opportunities for coordination in order to implement a human services plan and the job access and reverse commute (JARC) program funding application for Herkimer and Oneida Counties.

Oneida County Department of Planning, *Creating a Greenway in Oneida County: Part of the Mohawk River Corridor* (June 2008, updated November 2008) – This report details the greenway corridor concept for the Oneida County segment of the Mohawk River Trail. The report presents the existing conditions, significant nodes of interest, partnership opportunities, and methods of promoting the corridor.

National Park Service, *Erie Canalway National Heritage Corridor, Manifest for a 21st Century Canalway, Highlights of the Preservation and Management Plan (2006)* – This report outlines the preservation and management goals for the Erie Canalway National Heritage Corridor. The plan is intended to provide guidance to the Heritage Corridor Commission in order to provide support on the implementation of the goals laid out in the plan. The report received approval from the Governor of New York and the Secretary of the Interior and received an American Planning Association Award in 2008.

New York State Canal Corporation, *Who's on the Trail? The Canalway Trail User Count* (2006) – This report was put together by Parks and Trails New York in Albany for the New York State Canal Corporation. The report provides information on the number and types of trail users within Herkimer, Oneida, and Montgomery Counties. User counts took place in August and September of 2006 using volunteers by Parks and Trails New York. Within Oneida County, the report states that there are more bicyclists than joggers and walkers during the middle of the day in the town of Marcy. In addition, there are more joggers than bicyclists and walkers during the early evening in Rome. Throughout the three counties and the various survey points there were a similar number of bicyclists (43 percent) and walkers (35 percent), and a smaller number of joggers (20 percent). Of the trails surveyed, there were no wheelchair users, equestrians, or in line skaters observed.

New York State Department of Transportation, *New York State Bicycle Route 5* – This map includes the designated bicycle routes for the State of New York. The map details the road sections and towns along bicycle routes 5, 9, and 17. The map includes the existing routes listed above, as well as proposed bicycle routes, including bicycle routes 11, 19, 24, and 90.

Nonmotorized trail connections

Overview

Fort Stanwix National Monument has the potential to be a bicycle and pedestrian center for the city of Rome, nearby historic sites and attractions, and a valuable resource for Erie Canal corridor users. The park is well-positioned to serve bicycle and pedestrian trail users in addition to park visitors since its visitor center already provides public facilities, information, and access to several attractions. The nearby downtown area of Rome also contains amenities for visitors such as coffee shops, restaurants, and hotels. Many of the city's visitor amenities are concentrated along James Street, Dominick Street, Erie Boulevard, and Black River Boulevard, which are all a short distance from the park. By connecting to existing nonmotorized trails in the short term and coordinating with stakeholders and local planning agencies in the planning of trails in the long term, the park can attract visitors traveling along trail networks, provide recreational and educational opportunities for visitors, and increase public awareness of and access to nearby historic sites and attractions.

Purpose

The park's 2009 General Management Plan (GMP) recognizes that there exists little infrastructure for nonmotorized use. Nonmotorized use near the park is limited due to the roadway configuration, high vehicular speeds, and the numerous highway ramps and overpasses that do not accommodate pedestrians and bicyclists.[1] In light of these challenges, the park would like to increase nonmotorized access and visitation for three reasons: 1) to provide an alternative to driving and potentially attract more people to the park using several existing and planned nonmotorized facilities that can and should connect to the park; 2) for public health reasons, the city and county would like to see more nonmotorized activity in the city; and 3) all potential visitors do not have access to a vehicle, so a safe alternative to driving to the park should be provided. The purpose of this section is to identify and recommend ways to increase nonmotorized access to the park with a focus on bicycle and pedestrian trail connections. However, due to the small scope of this project, the recommendations should be considered preliminary; any alternatives presented in this section are not meant to substitute for a National Environmental Policy Act analysis of alternatives.

In order to encourage nonmotorized access, the park should pursue a number key connections, trail alignments, design changes, and signage additions that are described in this section. Note that the signage wayfinding section will focus on the location and details of signage for vehicular traffic and the pedestrian access and safety section will focus on safety and signage adjacent to the park.

By coordinating with local stakeholders such as the city of Rome; the New York State Canal Corporation; the New York State Office of Parks, Recreation, and Historic Preservation; the North Country Trail Association, the Erie Canalway National Heritage Corridor, and Oneida County, the park should influence new trail alignment and connections to the park. With new trail connections, the park can encourage bicycle and pedestrian access as an alternative to vehicular access to the park and the city of Rome. Trail connections should provide users with an enjoyable route, encourage people to be more physically active, and enable visitors who do not have access to a vehicle to visit the park. Trail connections would also increase public awareness of the park and other nearby recreational and historic sites within the region. Increasing park visitation and access to nearby sites lengthens visitors' time in Rome and the surrounding region, which improves the local economy.

Background

In addition to the park's efforts to increase nonmotorized awareness and visitation to the park, the Herkimer-Oneida County Long Range Transportation Plan includes bicycle and pedestrian

---

[1] National Park Service. Fort Stanwix National Monument 2009: Final General Management Plan and Environmental Impact Statement.

recommendations for the region.[1] The 2009 plan addresses the increase in public support for bicycle and pedestrian facilities in the region. The plan recognizes that several municipalities are not addressing bicycle and pedestrian facilities in their comprehensive plans, and that bicycle and pedestrian plans are typically not connected to other trail networks. The plan recognizes the importance of bicycle and pedestrian facilities and strives to make Herkimer and Oneida Counties a bicycle tourism destination. In addition, the plan recommends the need for programs that teach bicycle and pedestrian safety and the health benefits of nonmotorized travel. It recognizes the need for infrastructure improvements for roadway safety, the need for network connectivity, and the importance of signage and maintenance to create a bicycle- and pedestrian-friendly region.

The Herkimer-Oneida County Long Range Transportation Plan suggests that municipalities implement recommendations articulated in the Herkimer-Oneida County Bicycle and Pedestrian Plan to encourage pedestrian and bicycle use.[2] The Bicycle and Pedestrian Plan, updated in 2002, establishes a framework to create safe and attractive bicycle and pedestrian facilities, develop educational programs, and to provide interconnected bicycle and pedestrian systems throughout the region.

In addition to county level plans, Rome is working to improve its bicycle and pedestrian access and connectivity. The North County Trail Association is also working on a trail segment through Rome that will coincide with other proposed trail alignments. The nonmotorized trail infrastructure within Rome consists of sidewalks or short trail segments near the park. There are several proposed regional trails as well as additional connections that should improve access from trails and attractions to the park or from the park to nearby trails and attractions.

Existing and proposed trails

This section describes existing and recommended trail alignments and design in the region around the park (Figure 16) including the New York State Canalway Trail (Canalway Trail), Mohawk River Trail, city-owned multi-use trails, and New York Bicycle Route 5. The trails are described below in three parts: existing conditions; current county, city, or agency plans; and study team recommendations. In addition to the trail descriptions, Figure 17 and Figure 18 depict the recommended trail alignments. The study team's analysis and recommendations incorporate feedback from the park, city, county, and other stakeholder staff.

New York State Canalway Trail

Existing conditions

The Canalway Trail connects Albany to Buffalo with over 260 miles of existing trails.[3] However, there are gaps along the trail, including a gap in the city of Rome (Figure 15). Currently there is no designated or signed route in Rome.

Northwest of the city, the trail terminates three and a half miles from the park at the Rome Erie Canal Village (Figure 16). From here, the trail goes southwest towards Oneida. The trail is open to bicyclists, pedestrians, cross country skiers, equestrian riders, and snowmobilers. Southeast of Rome, the Canalway Trail terminates three miles from the park near the intersection of Martin Street and NY 365/NY 49 on Rome Oriskany Road/NY 69. Here, there is a small parking lot and Canalway Trail sign post that marks the start of the trail. The Canalway Trail from this location to the southeast towards Utica is made of crushed stone paving and is open to bicyclists, pedestrians, and cross country skiers in the winter.

---

[1] Herkimer-Oneida Counties Transportation Study. Long-Range Transportation Plan update 2010-2030.

[2] Herkimer-Oneida Counties Transportation Study. Bicycle and Pedestrian Advisory Committee (BAPAC). Herkimer-Oneida Counties Bicycle and Pedestrian Plan. 2002.

[3] New York State. Tourism and Recreation. http://www.canals.ny.gov/exvac/trail/index.html.

Current plans

The New York State Department of Transportation (NYSDOT) made $2.6 million available to the city of Rome and the Canal Corporation to develop the missing trail segment of the Canalway Trail in Rome; discussions for trail alignments are on-going. In June 2010, during a discussion with the city of Rome and the park, the Canal Corporation proposed a trail alignment that links Bellamy Harbor to the existing trail head and parking lot at NY69/Rome Oriskany Road. The Canal Corporation also identified available funding for additional trail improvement projects within the city. Planning meetings for trail alignments in Rome will continue between the city, the Canal Corporation, and local agency stakeholders.

The city of Rome's most recent proposal is outlined in Table 2. The proposed route does not include designated bicycle lanes or multi-use trails but should include signage to warn vehicles of bicycle traffic.

The study team believes that an on-street alignment is not ideal for visitors who are bicycling to or around the park, who may not be comfortable with no separation between them and moving traffic, especially if there is no or only a narrow shoulder along the street. The proposed connection from the trail at Erie Canal Village to the park has not been identified at this time; however, there are discussions within the city and county of several on-street alignments. Possible alignments include West Liberty Street, which ranges from two lanes of traffic with no shoulder to four lanes of traffic with parking on one side of the street. There is a narrow sidewalk along most of the north side of the road. The roadway is narrow for new or inexperienced bicyclists. Similarly, the connection along Erie Boulevard East also has four lanes of traffic and a narrow shoulder.

Recommendations

Several proposed routes in Rome are on-street without a designated bicycle lane. The study team recommends an alignment that connects the existing trails (at the Rome Erie Canal Village and at the trail head on NY69/Rome Oriskany Road) with off-street trails or designated bicycle lanes. The study team also recommends a connection from the existing Canalway trail to Oriskany Battlefield. Compared to a bicycle route that is completely on-street, a designated multi-use pedestrian and bicycle trail or a pedestrian sidewalk and designated bicycle lane would attract more users, encourage users to stay on the route, and improve trail user safety (Figure 21 shows bicyclists' preferences with regard to these types of facilities). Bicycle and pedestrian signage would improve wayfinding through Rome past the park; however, a designated bicycle and pedestrian trail would encourage more visitor and community trail use. An on-street alignment without a bicycle lane is not a recommended alternative. Table 1, Figure 17, and Figure 18 describe the recommended alignment.

The portion of the trail from Rome Erie Canal Village along National Grid Property to Muck Road is proposed by Herkimer and Oneida Counties.[1] If this route is not feasible due to existing ownership or liability concerns, an on-street connection with a bicycle lane should follow the roads near Rome State Wildlife Management Area including Muck Road and Seifert Road. The trail segment along South James Street should include a designated bicycle lane and sidewalk or a multi-use trail adjacent to the road. One lane of parking should be taken away or the sidewalk on one side of the road should be widened. The connection along South James Street is important in order to connect Canalway Trail users to the park.

In addition to connecting the existing segments of the Canalway Trail through Rome, the park would benefit from a connection to Oriskany Battlefield. This connection would provide a new opportunity for a park- or volunteer-guided tour on bicycle or a one way tour on foot with a shuttle ride back. The link would also provide an outdoor recreational option for visitors already traveling to the park. Alternatively, Thomas Road could be utilized to connect the Canalway Trail to Rome Oriskany Road; however, it is an additional mile south of the battlefield. This road would provide a connection between the trail and the battlefield using existing infrastructure.

---

[1] Herkimer-Oneida Counties Transportation Study *Herkimer and Oneida Counties (2007) Bicycling Atlas* (2007).

Table 2

Proposed and recommended routes for the Canalway Trail (from east to west)

Source: City of Rome and the Volpe Center

Proposed Route	Study Team's Recommended Route (Numbers are labeled in Figure 18)
• On-street from Rome Erie Canal Village to West Liberty Street • On-street adjacent to the park along North James Street • On-street along Erie Boulevard to Bellamy Park • On-street across the Erie Canal on Mill Street to Martin Street or NY 365/NY 49 • On-street to NY 69/Rome Oriskany Road to the existing Canalway trail parking lot off of Rome-Oriskany Road	1. Off-street from Rome Erie Canal Village along National Grid Property to Muck Road or on-street with bike lanes from the Canalway Trail, along Seifert Road to Muck Road. 2. On-street with bike lanes along Muck Road to South James Street 3. On-street with bike lanes or off-street along South James Street to North James Street 4. On-street with bike lanes along the park to East Dominick Street 5. Off-street along the Mohawk River along National Grid Property to Bellamy Park 6. On-street with bike lanes across the Erie Canal on Mill Street to Martin Street or NY 365/NY 49 7. On-street with bike lanes to NY 69/Rome Oriskany Road to the existing Canalway trail parking lot off of Rome-Oriskany Road 8. Off-street along the existing Erie Canalway Trail to a new spur trail that connects to Oriskany Battlefield (segment shown in Figure 17)

Figure 15

Erie Canalway Trail and canal navigation

Source: New York State Canal Corporation http://www.nyscanals.gov/maps/map5.html

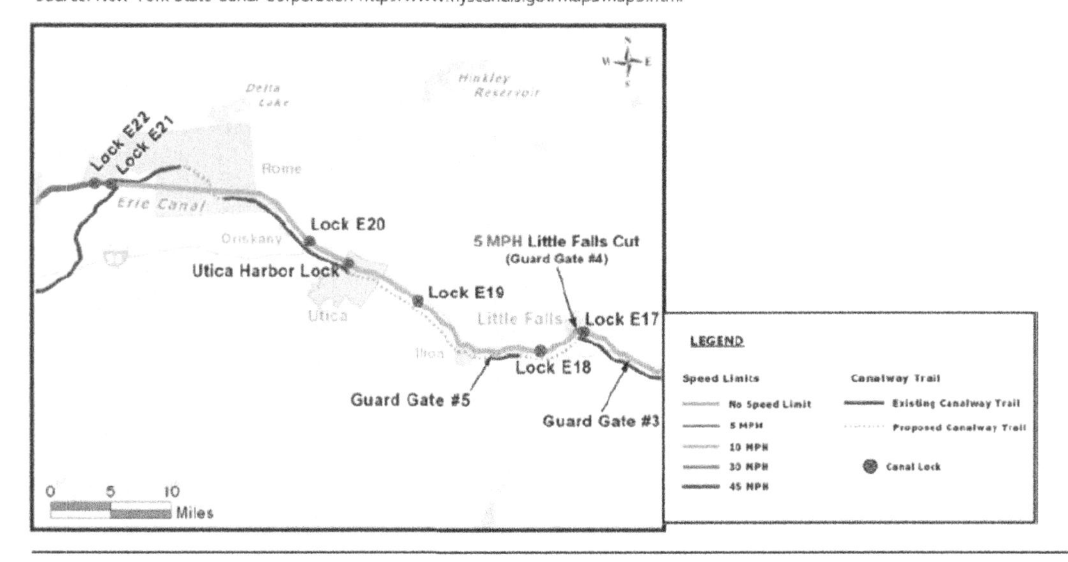

Figure 16
Existing trail network in the region
Source: Volpe Center, NPS, Herkimer and Oneida Counties

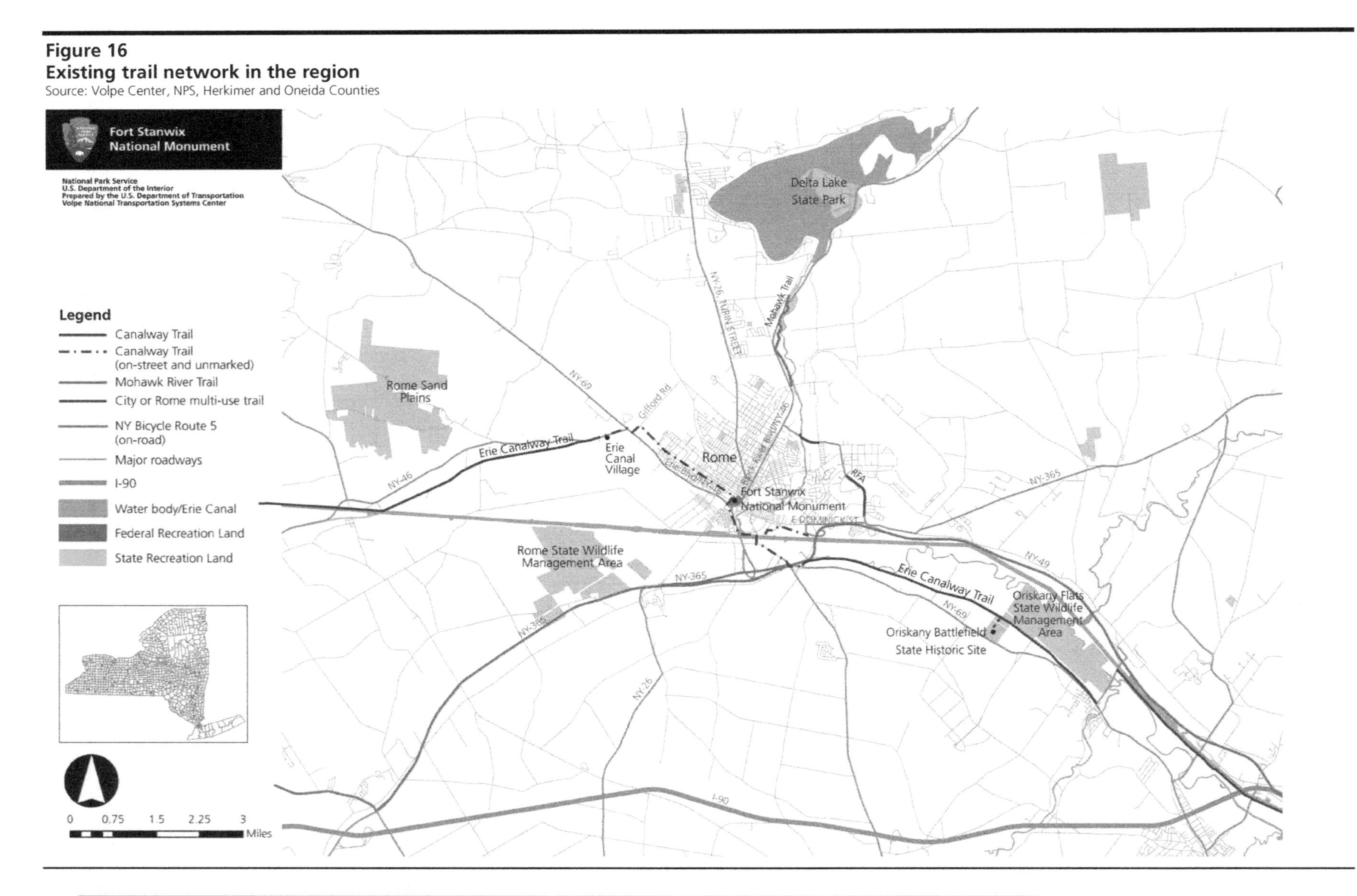

Figure 17

Trail alignment for region proposed by study team

Source: Volpe Center, NPS, Herkimer and Oneida Counties

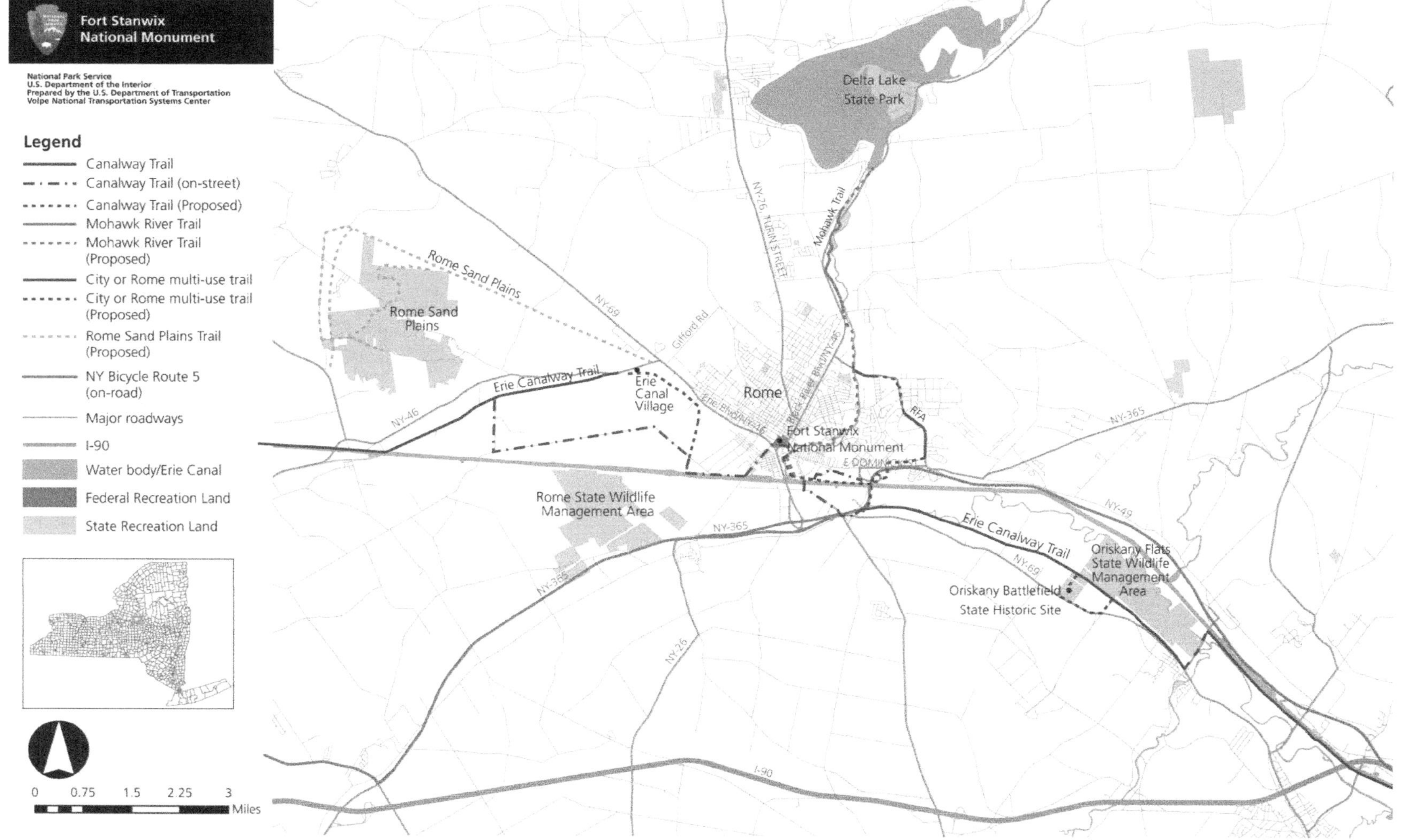

Figure 18
Trail alignment for Rome proposed by study team
Source: Volpe Center, NPS, Herkimer and Oneida Counties

Mohawk River Trail

Existing conditions

The city of Rome's parks department is currently planning the alignment of the Mohawk River Trail. When complete, the trail will link the Erie Canal at Bellamy Park to Delta Lake State Park with spurs into downtown Rome and to the park. The Mohawk River Trail will be one segment of the North Country Scenic Trail (for additional context and recommendations for the North Country Scenic Trail, see the North Country Trail section, below).

Current plans

Rome is considering first designating the Mohawk River Trail on-street without bicycle lanes between Bellamy Park and Fort Stanwix. The current proposal for the trail is outlined in Table 2. Erie Boulevard has high traffic speeds and receives heavy traffic entering and exiting the city. Originally, Rome considered trail alignments along the National Grid property adjacent to the Mohawk River and on-road segments along Mill Street or Harborway (Figure 19). For the short term, the city plans to designate on-street bicycle routes while seeking permission to design and construct an off-street trail along the Mohawk River. These routes are both shown in magenta in Figure 19.

Rome received funding from the New York State Environmental Protection Fund (EPF) in 2009 for improvements to Bellamy Park. In addition, the city is planning for several restoration and remediation projects near the canal and downtown area as part of the state's Brownfield Opportunities Area (BOA) program. These projects are currently on hold until more coordination takes place. The intention of the city's planned trail head and construction of trail user facilities at Pinti Field is to provide a connection to surrounding trails, the elementary school, and water activities.

Recommendations

The study team's recommended route is summarized in Table 3. The study team recommends a separate off-street trail on National Grid Property to connect Bellamy Park to the park. Although the process to secure this right of way and construct a new trail may take time, a designated multi-use trail that is separate from vehicular traffic will be safer and more enjoyable for trail users in the future. This multi-use trail will serve the local community as well as tourists who visit the Erie Canal and the city of Rome. The construction of a single multi-use trail along National Grid Property will be less expensive in the long run than an on-street trail designation now and a future off-street trail designation later. Additionally, there are interpretive opportunities on which the park and city can collaborate. The park should coordinate with the city of Rome to ensure that a well signed spur trail connects from the Mohawk River Trail to the park along East Dominick Street (Figure 17 and Figure 18).

The study team believes that trail head amenities that are proposed for Pinti Field would be a duplication of amenities already available at Fort Stanwix. New boat launch and elementary school facilities might still be appropriate at Pinti Field; however, the study team recognizes the importance of a single regional trailhead. The park's existing facilities, visitor information, and park staff can serve park visitors, trail users, and function as the regional trailhead.

Table 3
Proposed and recommended routes for the Mohawk River Trail (from north to south)
Source: Volpe Center

Proposed Route	Study Team's Recommended Route
• Combination of off-street and on-street from Delta Lake State Park to Fort Stanwix • Trail head at Pinti Field • On-street (short term) and off street (long term) from Fort Stanwix along Erie Boulevard to Bellamy Park (Canalway Trail alignment, see Figure 19)	• Off-street from Delta Lake State Park to Fort Stanwix • Use of Fort Stanwix as a trail head • On-street with bike lanes from the park on East Dominick Street and off-street along the Mohawk River to Bellamy Park

Figure 19
Mohawk River Trail concepts
Source: city of Rome and NPS

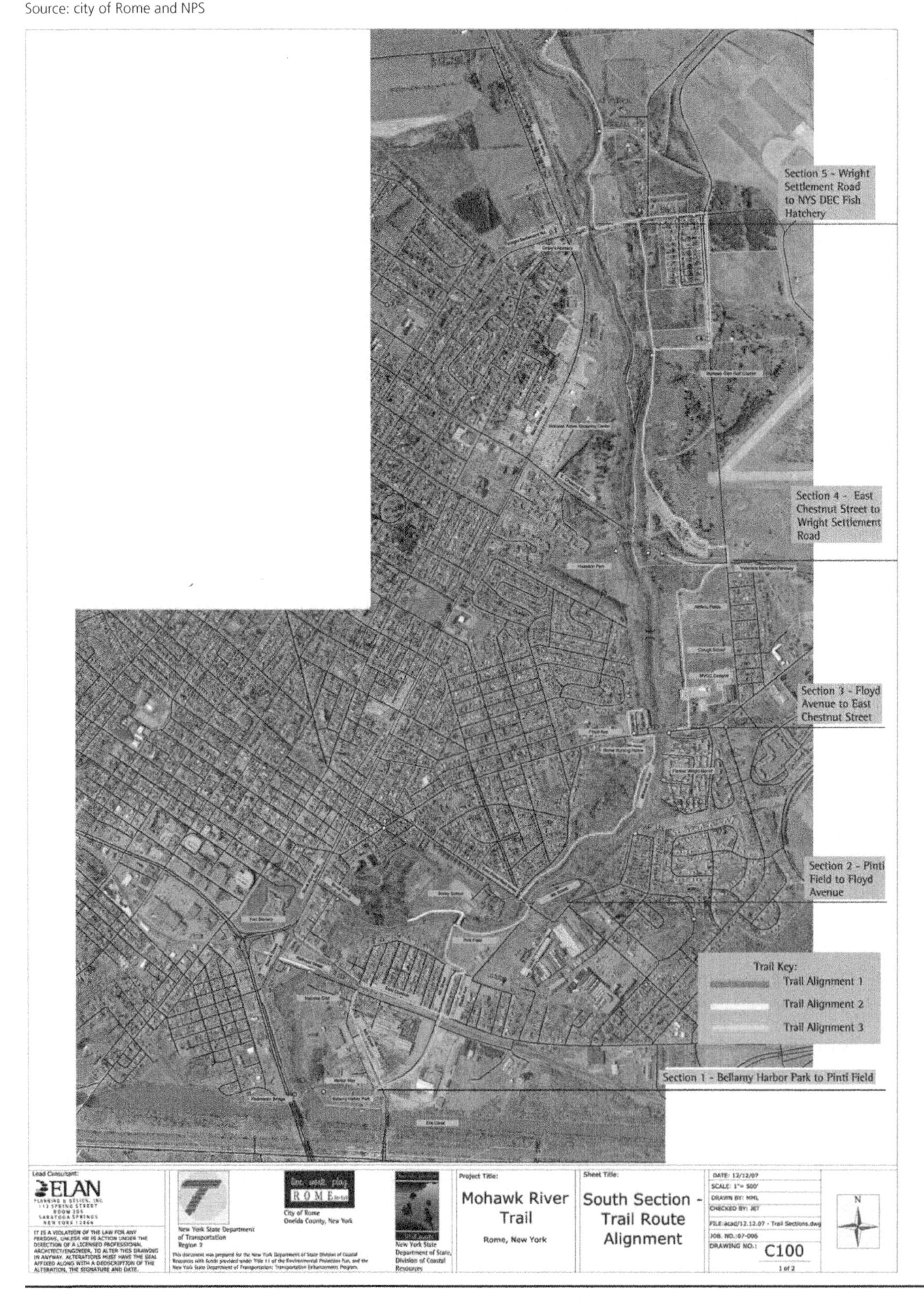

The Great Carry Trail (De O Wain Sta)

Existing conditions

The Great Carry Trail is the historical connection between the Mohawk River to Wood Creek. This trail made it possible for fur traders to transport goods from the Hudson River to the Great Lakes. The trail was used by Seneca, Cayuga, Onondaga, Oneida, Mohawk, and Tuscarora as well as the English, Canadians, and French. The construction of Fort Stanwix in the 18th century helped to protect the route between the water bodies for the transport of goods. The trail varied from one to six miles depending on the time of year and location of the waterways. Today, the approximate trail alignment crosses near Bellamy Park, the park, and City Hall.

Current plans

There is currently no designated trail to commemorate this connection due to the trail's fluctuation in length and alignment through the downtown.

Recommendations

If stakeholders begin to pursue planning such a trail, the connection should be easy to navigate and safe for all users in order to encourage visitors to walk or bicycle along this route. The study team recommends that the trail between Bellamy Park and Fort Stanwix should share an alignment with the Mohawk River Trail and provide interpretive education to trail users. As other segments of the Canalway Trail are planned near the park, there would be further opportunity for the Great Carry Trail to share an alignment. Alternatively, the Great Carry Trail could follow existing sidewalks if the park and stakeholders work to create a self-guided tour. Park staff should coordinate with the Oneida Indian Nation on the trail location and its interpretation.

City of Rome multi-use trails

Existing conditions

Rome has a one mile multi-use trail between Rome Free Academy High School (RFA) and Griffiss Air Force Base. This trail is separate from vehicular traffic and accommodates both pedestrian and bicycle use. No other multi-use trails, in addition to those mentioned earlier, exist or are planned by the city of Rome at this time.

Current plans and recommendations

The study team recommends that the park should continue to coordinate with the city of Rome as new multi-use trails are planned or if the city extends the RFA trail. As more trails are built within the city, the park has the opportunity to encourage more visitors to use trails to access the park and nearby destinations. Trails should serve visitors who are unfamiliar with the area as well as local residents who use these routes to commute or exercise.

New York Bicycle Route 5

Existing conditions

Bicycle Route 5 follows NY 365 and NY 49 near Rome. This route is recommended by NYSDOT for experienced bicyclists traveling long distances. This on-road bicycle route is 365 miles long and connects Niagara Falls and Albany. The route contains special signage and a striped shoulder that separates vehicular and bicycle traffic. The shoulder width varies between four and six feet, which is appropriate for experienced bicyclists.

Current plans and recommendations

This bicycle route is less than two miles south of the park; however, the connection from the route to the park is not clear to bicyclists. Signage and a designated connection between the bicycle route and the park would encourage Route 5 users to visit the park. The study team recommends that park staff should work with NYSDOT to construct new signs along Route 5 at the Oriskany exit to instruct bicyclists who are interested in visiting the park to follow Lampher Road to the proposed Canalway Trail and to the park.

North Country National Scenic Trail

Existing conditions

The North Country National Scenic Trail travels through North Dakota, Minnesota, Wisconsin, Michigan, Ohio, Pennsylvania, and New York. There are approximately 2,200 usable miles of the total 4,600 mile long trail (Figure 20).[1]

Current plans

The North Country Trail Association works with local agencies and stakeholders to designate segments of trail along a multi-state route. The Association is proposing to use the Mohawk River Trail that follows the Old Black River Canal towards Boonville where it enters the Adirondack State Park.[2] Most of the trail route is a hiking trail with additional trail users determined by local agencies; bicycling is permitted on several trail segments. Snowmobiling may be a future use on the trail since it is a historic use in some trail segments. If snowmobiling is permitted along the trail in the future, the trail should be wide enough to accommodate this use in addition to other winter uses such as cross country skiing.

Recommendations

The study team recommends that the park work with the Association to convey the importance of off-street bicycle and pedestrian routes near the park and to advocate for locating the Mohawk Trail along the study team's preferred route. Further, the park should promote the visitor center as a trailhead or rest area for trail users. Park staff should assist trail users and direct visitors to amenities within the park and city.

Rome Sand Plains Trail

Existing conditions

The Rome Sand Plains Management Team includes representatives from the New York State Department of Conservation, the Nature Conservancy, Oneida County, NYSDOT, the city of Rome, and several other local organizations. Together, these agencies manage the Rome Sand Plains Resource Management Area. Designated in 1997, the management area protects inland pine barrens, sand dunes, and wetlands. There are two hiking trails, the Wood Creek Trail (1.2 miles) and the Sand Dune Trail (1.9 miles).

Current plans

The partner organizations responsible for the management of the area are preparing a management plan in order to protect, enhance, and provide additional recreational opportunities for the area.

Recommendations

The study team recommends that the park staff work with the city of Rome and Rome Sand Plains staff to construct a trail connection between Rome Sand Plains and the park. This connection would provide visitors to the park with additional connections to recreation opportunities. The connection from the park to Rome Sand Plains should follow the same alignment as the study team's preferred alignment for the Canalway Trail (Table 1, Figure 17, and Figure 18) to the intersection of Rome New London Road/NY46/NY49 and Erie Boulevard West/NY46/NY49/NY69. At this location, Herkimer and Oneida Counties propose an off-street alignment along the National Grid property to Rome Sand Plains.[3]

---

[1] Phone conversation with Albert Larmann, Central New York Coordinator for the North Country National Scenic Trail. Participants included Mr. Larmann and study team. May 24, 2010.

[2] National Park Service, North Country National Scenic Trail Association, and New York State Department of Conservation. North Country National Scenic Trail Final Adirondack Park Trail Plan/ Final Generic Environmental Impact Statement. July 2008.

[3] Herkimer-Oneida Counties Transportation Study *Herkimer and Oneida Counties (2007) Bicycling Atlas* (2007).

Figure 20
North Country National Scenic Trail
Source: North Country National Scenic Trail http://www.nps.gov/noco/index.htm

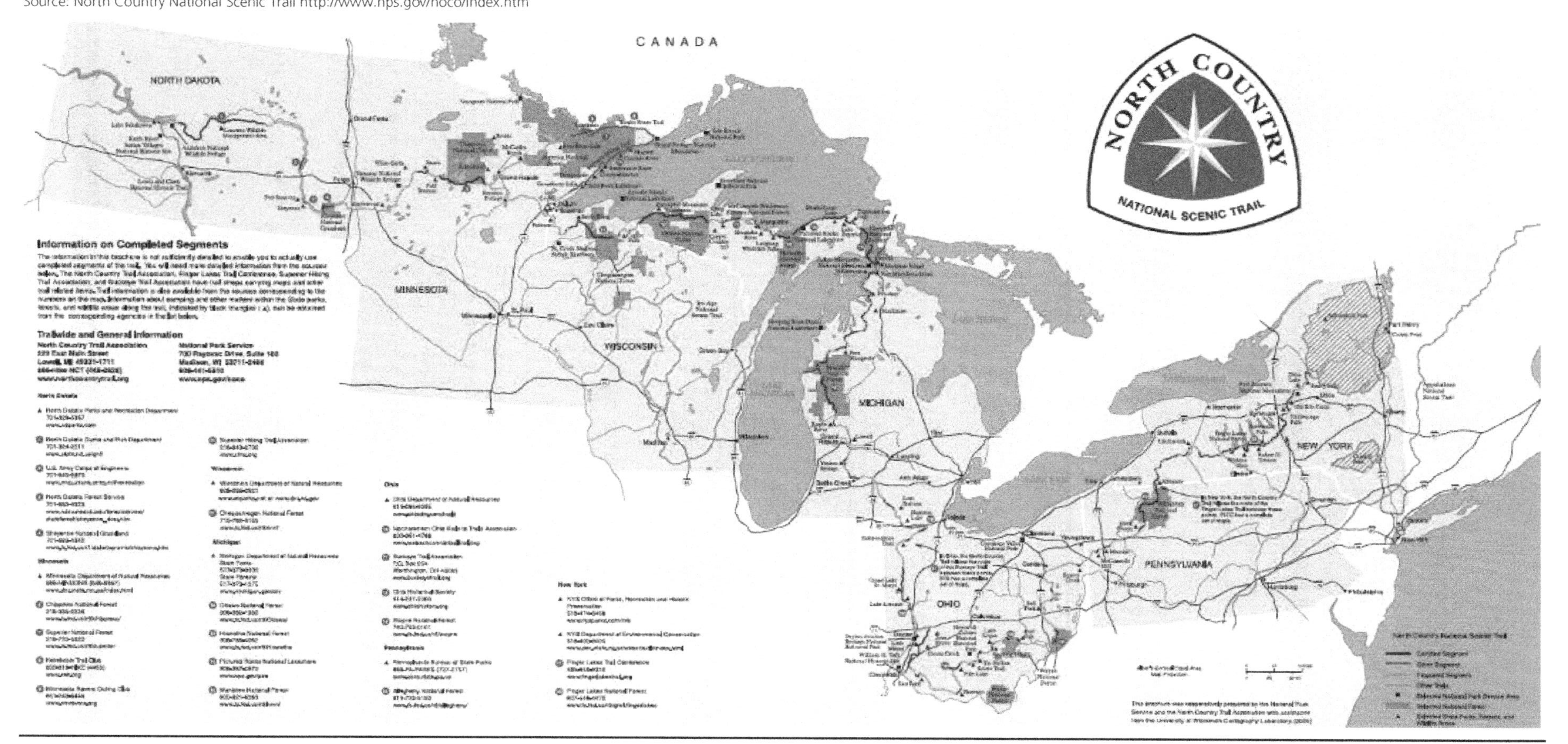

Trail design and visitor experience

New on-street bicycle lanes, multi-use trails, and pedestrian trails should accommodate both existing and future user groups, which include children, families, and senior citizens. Long distance bicycle routes and popular commute routes accommodate faster bicycle speeds and experienced bicyclists; these routes can run along traffic or on narrow roadways. Recreational, multi-use, or short distance trails should accommodate group trips and inexperienced user groups. These types of routes should have a wider clearance and, when on-street, shoulders to allow for a variety of users. Each of these facilities has advantages and disadvantages, discussed below, as well as spatial and technical requirements depending on the right-of-way and existing infrastructure.

Bicyclists generally prefer bicycling on off-street paths and multi-use trails. Figure 21 shows the results of a 2007 study conducted in New York City.[1] Over 75 percent of bicyclists surveyed stated that they prefer to ride on off-street paths compared to on-street. Nearly two-thirds of bicyclists surveyed also said that they will detour to a longer route to take an off-street trail.

Figure 21
Bicyclist preferences in New York City.
Source: The New York City Bicycle Survey, http://www.nyc.gov/html/dcp/pdf/transportation/bike_survey.pdf

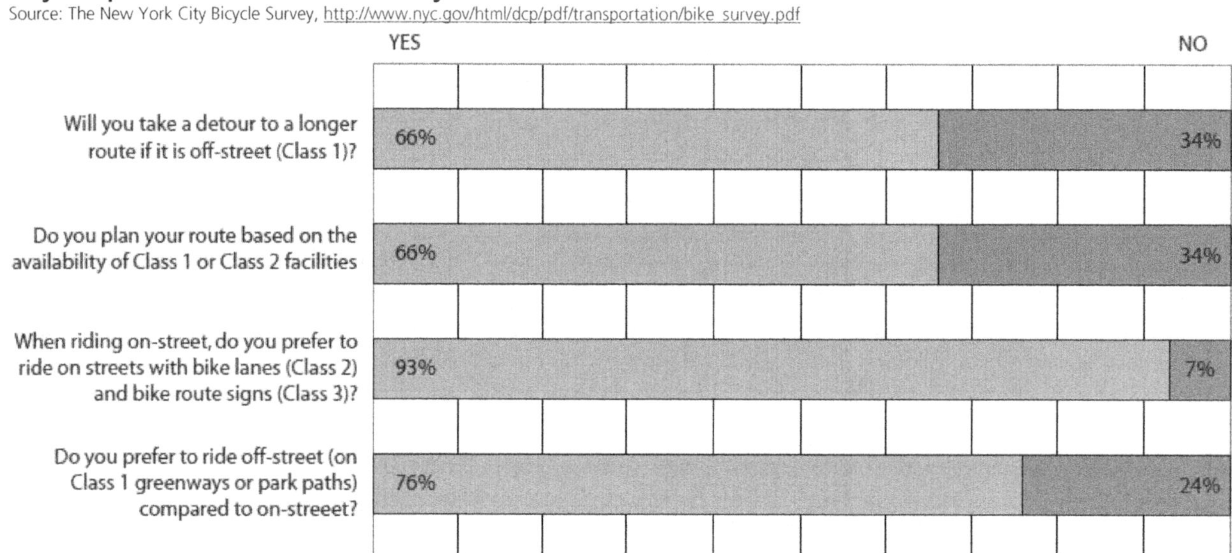

New trail alignments and design should consider existing traffic speeds, adjacent parking, existing and future trail users, and trail connections. Bicycle routes that are adjacent to heavy traffic or parked cars should include signage, lane markings, and as much separation from traffic as possible so that trail users are protected from vehicular traffic. All trails should meet Americans with Disabilities Act guidelines and follow the American Association of State Highway and Transportation Officials and state regulations.[2] Trails and bicycle routes should be signed so that users know the distance between attractions. In addition, signage that warns vehicles of nearby nonmotorized use will improve safety for trail users.

Some of the trail design options include the following (Figure 22):

---

[1] City of New York. The New York City Bicycle City. http://www.nyc.gov/html/dcp/pdf/transportation/bike_survey.pdf, 2007.

[2] U.S. Department of Transportation: Federal Highway Administration. *Manuals and Guides for Trail Design, Construction, Maintenance, and Operation, and for Signs*. http://www.fhwa.dot.gov/environment/rectrails/manuals.htm#aashto

- On-street bicycle lanes are adequate for experienced or long distance bicyclists rather than for group travel or inexperienced bicyclists, since on-street bicycle lanes can increase interaction between bicyclists and vehicular traffic. Installation/striping of bicycle lanes requires a roadway that is wide enough to accommodate designated bicycle lanes. A bicycle route without a designated bicycle lane would not be as safe for bicyclists as a trail with a designated bicycle lane.
- Multi-use trails, grade and distance separated from a vehicular roadway, are adequate for bicycle and pedestrian use and are safer for trail users than on-street facilities. Multi-use trails require a wider right of way than on-street bicycle lanes and often require new construction, which can make them more expensive.
- Pedestrian trails or sidewalks are typically for pedestrians and slow bicycle traffic. Trail paving varies and can include materials such as asphalt, concrete, and decomposed granite. Pedestrian trails are appropriate when there are on-street designated bicycle lanes.

Figure 22
Bicycle and pedestrian trails: on-street bicycle lane (left), multi-use trail (center) and pedestrian trail (right).
Source: www.pedbikeimages.org/ElaineNelson, www.pedbikeimages.org/DanBurden (center and right)

Special events

There are several special events along the Erie Canal that necessitate a safe and well-designated connection from the Canal and Canalway Trail to the park. An annual event that takes place at the park includes Cycling the Erie Canal, which takes place in July and includes a stop at the park where bicyclists camp overnight on the lawn (Figure 23). The event is designed to increase awareness to the Erie Canal and the recreational opportunities associated with the canal. The Canalway Trail Celebration is in June. Cities and organizations coordinate activities along the Erie Canal during this event. The Canal Splash, in August, is a series of events that celebrate recreation along the canal. The Rome Rotary Club sponsors Canalfest on the first weekend in August at Bellamy Park. Canalfest features run and kayak races and provides music, food, and fireworks.

Figure 23
Cycling the Canal stopover at Fort Stanwix
Source: John Dimura, Canal Corporation

Bicycle tourism

Several of the special events along the Erie Canal focus on trail and canal recreation. To improve park awareness and attract more visitors to Rome, park staff should work with the city of Rome and the Rome Area Chamber of Commerce to attract bicycle tourists and encourage them to use accommodations and amenities in Rome and at Fort Stanwix. According to a recent publication from the Erie Canalway National Heritage Corridor, New York State Canal Corporation, and New York Parks & Trails, bicycle tourism is increasing in popularity and there are unique opportunities to enhance tourism in towns along the Erie Canal.[1] The report finds that bicycle tourists spend a significant amount of money on food and accommodations and bicycling is already a popular sport along the canal. In addition, bicycle tourists seek locations based on the difficulty and length of ride, support and services along the route, and nearby attractions. The report recommends that business owners, restaurants, and lodging establishments become more bicycle friendly as more bicyclists come through town. Rome businesses could become more bicycle friendly by providing bicycle amenities including facilities, local information, bicycle parking, water, and food.

There is an opportunity for the park to expand interpretive programming to include bicycle or walking tours of the area surrounding the park.[2] With the construction of bicycle and pedestrian connections, park staff or volunteers could lead visitors from the park to nearby destinations such as Bellamy Park, the Erie Canal, Oriskany Battlefield, Erie Canal Village, Downtown Rome, or Delta Lake State Park. The park could also provide self-guided tours with maps and wayfinding signs to these sites. These tours would encourage outdoor recreation, educate the public of nearby sites, and encourage longer visits to the region. In the winter, tours from the park could be taken on cross-country skis or snowshoes.

Stakeholders, coordination, and partnerships

Park staff should continue to participate in city and regional trail planning discussions in order to convey the importance of connecting nearby trails to the park and promoting trail infrastructure that is safe for all users. Contact with the following agencies should continue.

---

[1] Erie Canalway National Heritage Corridor, New York State Canalway, and New York Parks & Trails "Bicyclists Bring Business: A Guide for Attracting Bicyclists to New York's Canal Communities," 2010.
[2] This concept came out of a conference call with Hannah Blake, from the Erie Canalway National Heritage Corridor, who reported that Saratoga National Historical Park provides a limited number of bicycle tours of the battlefield.

- City of Rome: Diane Shoemaker (dshoemaker@romecitygov.com) and Matthew Andrews (mandrews@romecitygov.com), responsible for city of Rome planning, city trails, and Bellamy Park planning. Tel: (315) 339-7608.
- Erie Canalway National Heritage Corridor: Hannah Blake (Hannah_Blake@nps.gov), Director of Planning and Heritage Development. Tel: (518) 237-7000, ext. 202.
- NYS Canal Corporation: John Dimura (John.Dimura@thruway.state.ny.us), Trails Director. Tel: (518) 436-3034. Agency website: http://www.nyscanals.gov/exvac/trail/index.html
- North Country Trail Association: Albert Larmann (aflarmann@msn.com), Central New York Coordinator for the North Country National Scenic Trail. Tel: 315-697-3387.
- NYS Office of Parks, Recreation, and Historic Preservation http://www.nysparks.state.ny.us/recreation/trails/
- NYSDOT bicycle coordinator: Paul Evans (pevans@dot.state.ny.us) Tel: (315) 793-2433 https://www.nysdot.gov/portal/page/portal/divisions/operating/opdm/local-programs-bureau/biking
- Rome Area Chamber of Commerce: William Guglielmo, (wkg@romechamber.com), President Tel: (315) 337-1700.

The park would benefit from coordination with other agencies such as Herkimer and Oneida Counties, the city of Rome, and the Canal Corporation to conduct trail user surveys along the Canalway Trail and on surrounding bicycle and pedestrian routes near Rome. These data would clarify what nonmotorized modes of travel are most common along the corridor, if trail users are visiting the park in addition to other recreation or historic sites, and if trail users are currently aware of the park's location.

The park could also coordinate with other agencies to install and locate wayfinding, trail information, and infrastructure and to plan events along the canal and through the city. The construction of a link trail from the Canalway Trail to Oriskany would benefit from the park coordinating with the Canal Corporation as well as the NYS Office of Parks, Recreation, and Historic Preservation. Collaboration with these partners can improve mobility in the region and promote tourism around canal and trail activities.

Conclusion

The park can encourage nonmotorized transportation and access to the park by working with local and state agencies to create connections to existing and planned nearby trails. New ways to access the park will provide users with an alternative to driving when visiting the park. Visitors will be more likely to use alternate routes if the trails are clearly designated and safe for trail users. Nonmotorized trail connections would increase visitation to and between nearby historic sites such as Oriskany Battlefield and Rome Canal Village and offer visitors new recreational opportunities.

Short term coordination should occur with the city of Rome and the Canal Corporation in order to create a connection within Rome for the Canalway Trail. The park should convey the importance of safe and clearly marked bicycle and pedestrian routes in order to encourage visitation and recreational options. Trail plans should recognize existing user groups and plan for future trail users including groups of inexperienced users and groups. Long term coordination should continue with Rome, the Erie Canalway National Heritage Corridor, and the North Country Trail Association in order to ensure safe trail alignments and connections to the park.

Signage and wayfinding

Overview

With additional signage, Fort Stanwix National Monument can achieve three of its expressed goals: improve visitor access, increase park visibility, and attract more visitors. To help achieve these ends, this section of the report inventories and evaluates existing signage and identifies opportunities for new vehicular directional signs within a 15-mile radius of the park. The section also summarizes the process, partners, and approximate cost of new signs. To install motorist wayfinding signs, the park should pursue funding and sign permits through the city of Rome, the New York Thruway Authority (NY Thruway), and the New York State Department of Transportation (NYSDOT).

Existing signage is often inconsistent, small, and relatively uninformative for motorists. New signage along major interstates and local corridors will improve visitors' experience of traveling to and arriving at the park and ensure that first time visitors and tour bus drivers are able to more easily navigate to the park. New signage near the park will also improve park visibility to motorists who are already in Rome. All new signs should inform regional motorists of the park's distance and direction.

Existing wayfinding

Currently, there are 17 signs for the park within a 15-mile radius of the park. This includes six blue and white signs, nine brown and white historic site signs, and two National Park Service (NPS) signs (Figure 24). A list of existing wayfinding or directional signs to the park (Tables 4 and 5) includes the route (see description of routes in next section), map number (map#-sign#), road name and intersecting road, sign text, jurisdiction, and description. Signs are grouped into two categories and two maps: signs along the NY Thruway (Interstate 90) on map 1 (Figure 25) and signs near Rome on map 2 (Figure 26).

Figure 24
Sign types: Sign along I-90 (left), Historic Site sign (center), and National Park Service sign (right)
Source: Volpe Center

Routes to Fort Stanwix

The tables (Tables 4 and 5) and maps (Figures 25 and 26) show the sign locations as well as the routes to the park. These routes follow major corridors from surrounding recreation areas, historic sites, cities, towns, and highways to the park. They represent the shortest routes from various locations to the park and were used by the study team to consider the best locations for new signs. The study team incorporated park staff input, site observations, and maps to determine the location of these routes. These routes are:

- Route A: Syracuse/I-90/NY365
- Route B: Rome Sand Plains/NY46/NY49
- Route C: Taberg/NY69
- Route D: West Leyden/NY26
- Route E: Boonville/NY46
- Route F: Holland Patent/Prospect/NY365
- Route G: Herkimer/NY49
- Route H: Oriskany/NY69
- Route I: Albany/NY69/I-90

Table 4
Signs on New York State Thruway (I-90) (Map 1)
Source: Volpe Center

Route	Map	Sign #	Road name	Intersection	Sign Text	Jurisdiction	Description
I	1	1	Exit 32 westbound	I-90	Fort Stanwix National Monument	Thruway Authority	Small blue sign with white text mounted on a large sign with other small signs
A	1	2	Exit 33 westbound	I-90	Fort Stanwix National Monument	Thruway Authority	Small blue sign with white text mounted on a large sign with other small signs
A	1	3	Exit 33 eastbound	I-90	Fort Stanwix National Monument	Thruway Authority	Small blue sign with white text mounted on a large sign with other small signs
I	1	4	Exit 32 eastbound	I-90	Fort Stanwix National Monument	Thruway Authority	Small blue sign with white text mounted on a large sign with other small signs
A	1	5	NY 365, near exit 33	NY 31	Fort Stanwix National Monument	Thruway Authority/ NYSDOT	Large blue and white sign, off NY 365
I	1	6	Cider Street	NY 233	Fort Stanwix National Monument	Thruway Authority	Small blue sign with white text mounted on a large sign with other small signs

Table 5
Signs in Rome (Map 2)
Source: Volpe Center

Route	Map	Sign #	Road name	Intersection	Sign Text	Jurisdiction	Description
G, H, I	2	1	NY 365/69/49	NY 26N/49W/ NY69W/ Erie Boulevard	Fort Stanwix Historic Site	City of Rome/ NYSDOT	Brown and white square sign
A	2	2	NY 365/26	NY 49	Fort Stanwix Historic Site	City of Rome/ NYSDOT	Brown and white square sign
H, I	2	3	Martin Street	NY 365	Fort Stanwix Historic Site	City of Rome/ NYSDOT	Brown and white square sign with left arrow
C	2	4	Erie Boulevard	north of NY 49/ 46 intersection	Fort Stanwix Historic Site	City of Rome/ NYSDOT	Brown and white square sign with straight arrow
B	2	5	NY 49	NY 49/46/69	Fort Stanwix Historic Site	City of Rome/ NYSDOT	Brown and white square sign with right arrow
B,C	2	6	Erie Boulevard	Gifford Road	Fort Stanwix Historic Site	City of Rome/ NYSDOT	Brown and white square sign with straight arrow
B,C	2	7	Erie Boulevard	South Levitt Street	Fort Stanwix Historic Site	City of Rome/ NYSDOT	Brown and white square sign with straight arrow
B,C	2	8	Erie Boulevard	Near North James Street	Fort Stanwix Historic Site	City of Rome/ NYSDOT	Brown and white square sign with left lane sign below
A,G,H,I	2	9	Erie Boulevard/NY 26/NY49 /NY 68	Black River Boulevard	Fort Stanwix National Monument	City of Rome/ NYSDOT	Brown and white square sign with straight arrow
E	2	10	Black River Boulevard	East Dominick Street	Fort Stanwix National Monument	NPS	Large brown and white sign with NPS arrow
F,G	2	11	Black River Boulevard	NY 26N/49W/ NY69W/Erie Boulevard	Fort Stanwix National Monument Parking	NPS	Large brown and white sign with arrow for parking

Figure 25

Map 1 - Signs along the NY Thruway (I-90)

Source: Volpe Center

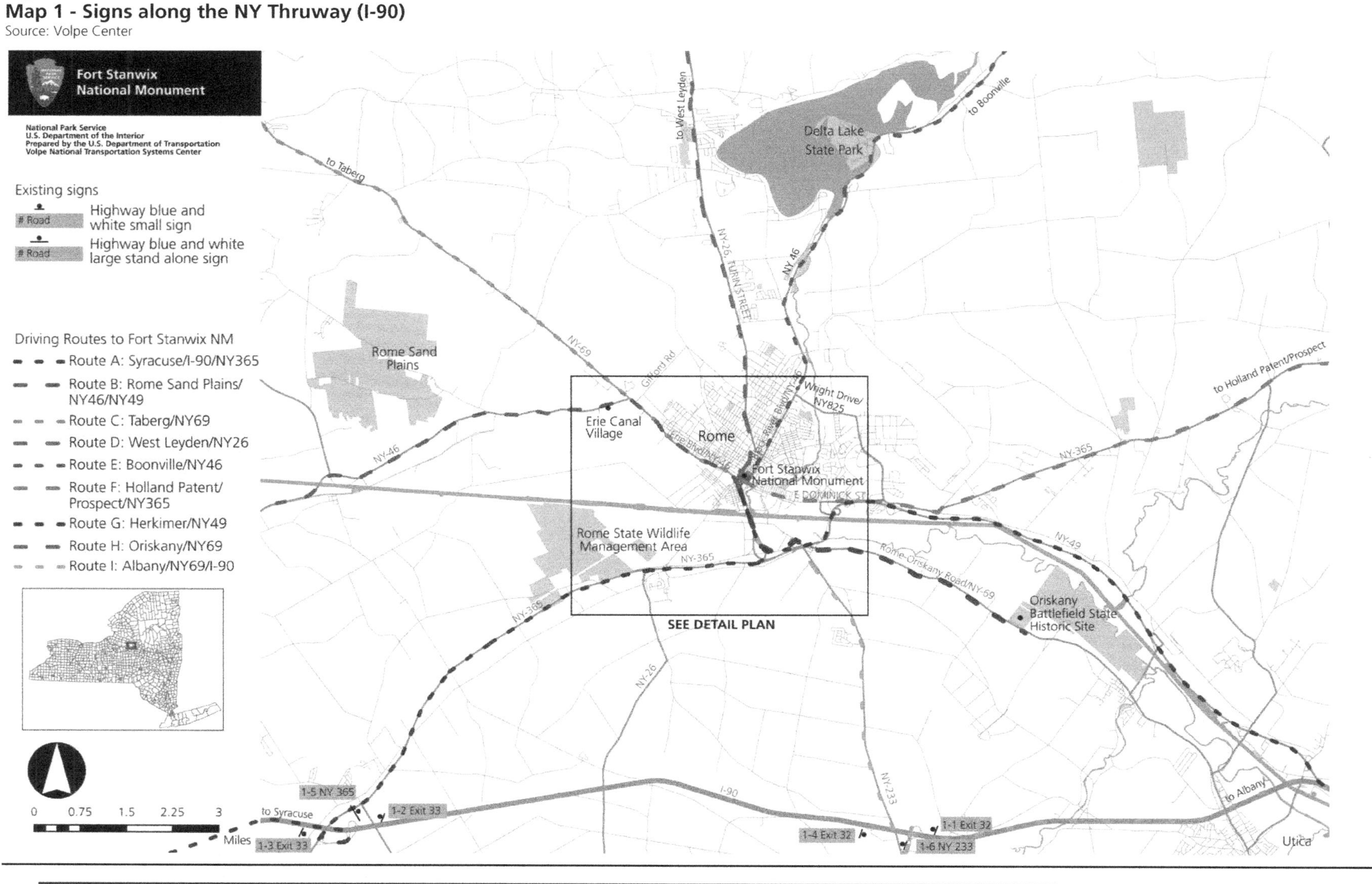

Figure 26
Map 2 - Signs in Rome
Source: Volpe Center

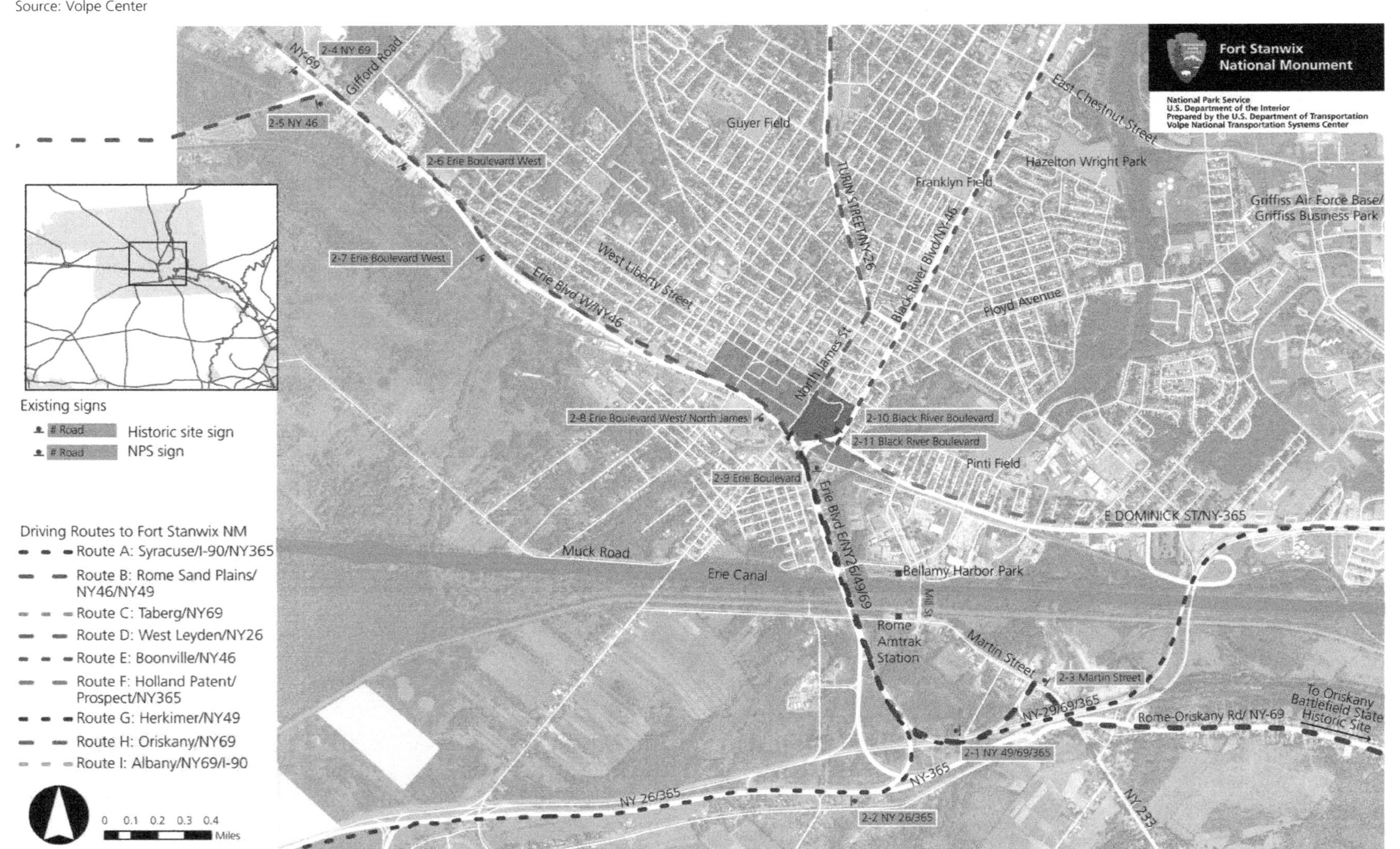

Existing sign locations and descriptions

There are four signs along the NY Thruway. There are signs for the park at exits 32 and 33 for westbound motorists and signs for the park at exits 33 and 32 for eastbound motorists (Figure 27). These four signs are under the NY Thruway Authority jurisdiction. The signs are blue and white and mounted on large blue signs that typically contain other signs for hotels, gas stations, and recreation sites. The signs for the park are approximately two feet tall by three feet wide with the words "Fort Stanwix Nat'l Monument" and the NPS arrowhead (Figure 24, left).

Figure 27
Signs on NY Thruway: exit 32 and 33 westbound (left), exit 33 and 32 eastbound (right)

In addition to the signs along the NY Thruway, there are two blue and white signs just off the NY Thruway along the route to the park. There is one sign for northbound motorists on NY 365 and one sign for northbound motorists on NY23 (Figure 28). The sign on NY 365 contains large text that is easy for motorists to read and the NPS arrowhead. Both signs include mileage and directional information.

Figure 28
Signs off of NY Thruway: exit 33 on NY 365 (left), exit 32 on NY 233 (right)
Source: Volpe Center

There are eleven existing signs along NYSDOT roadways and city of Rome roads. These signs are typically brown and white "Historic Site" signs (Figure 24, center). These signs are typically installed and managed in coordination with the NYSDOT and with the city of Rome when located within the city. There are three of these signs south of Rome on Erie Boulevard/NY 26/NY49 /NY 69 for northbound motorists. One sign is on NY 49/NY 69/NY 365 for northwest bound traffic and a second sign is on NY 26/NY 365 for northeast bound motorists. In addition there is a sign for northbound motorists on Martin Street going towards NY365 (Figure 29).

Figure 29
Signs near Rome
Source: Volpe Center

In Rome, there are five brown and white "Historic Site" signs with arrows for motorists traveling southeast into downtown Rome along Erie Boulevard West/NY46/NY49/NY69 (Figure 30). The first park sign is near the intersection of NY46/ NY49 and Erie Boulevard West/NY46/NY49/NY69. The sign directs motorists to the two historic sites in the area: Fort Stanwix and Erie Canal Village. The second sign is on NY 46/NY 49 before the intersection with Erie Boulevard West/NY46/NY49/NY69. The next three signs are along Erie Boulevard West/NY46/NY49/NY69 near Gifford Road, South Levitt Street, and North James Street.

Figure 30

There is a sign for northbound motorists on Erie Boulevard/NY 26/NY49/NY 69 (Figure 31) near the intersection with Black River Boulevard/NY46. There are also two NPS signs within the park's boundary for southwest bound motorists on Black River Boulevard/NY46 before the intersection with Erie Boulevard/NY 26/NY49/NY 69 (also Figure 31). The first sign includes the park name and direction, the second sign contains information for where motorists should park.

Figure 31
Signs near Fort Stanwix on Erie Boulevard/NY 26/NY49 /NY 69 and Black River Boulevard/NY46
Source: Volpe Center

Existing sign evaluation

The blue and white signs along the NY Thruway direct motorists to what exits to use to access the park. These signs are small, contain small text, and contain no mileage information. Accordingly, these signs are difficult to read at high speeds, which are common along the interstate. The larger blue and white signs that contain mileage information (as in Figure 28, left) are more visible and informative to motorists. The brown and white "Historic Site" signs are along some of the major corridors leading into Rome. These signs are less visible and informative when motorists are traveling at faster speeds on highways (left and center in Figure 29) since like the small blue and white signs, the text with the site name is small. These signs are more effective for motorists traveling at slow speeds (such as along the roadways pictured in Figure 30). A sign like the one found on the NY Thruway for the Erie Canalway National Heritage Corridor (Figure 32) is easy to read and easy to distinguish from other signs.

Figure 32
Existing Erie Canalway NHC sign
Source: Volpe Center

New sign recommendations

Well-placed and consistent signage will improve wayfinding at intersections and inform motorists where to make a turn to access the park. The major travel routes, destinations, and population centers are important factors in determining the location for new signs. The busiest roadways near the park, according to annual average daily traffic (AADT) data are the following:

- NY Thruway Exit 33 – 34,358
- NY Thruway Exit 31 – 23,436
- Black River Boulevard – 22,711
- Erie Boulevard West – 22,427
- NY 365 – 20,392
- NY 26/49/69 – 17,917
- Erie Boulevard East – 12,504

Sign selection

To determine the location for new signs, the study team documented existing sign locations and driving routes to identify gaps in sign locations. While some routes contain several existing signs, other routes contain no signs. The study team noted where motorists need to be informed of major turns or exits.

The study team's analysis results in 17 new recommended signs. Sign text should direct motorists to the park and provide the number of miles to the park. A directional arrow should be provided as appropriate. The study team recommends that all new signs are consistent in color, size, and be stand alone, similar to the sign in Figure 32. Where possible, the study team recommends brown and white signs with the NPS arrowhead because they are the most recognizable and are consistent with NPS signs in other parts of the country.

Sign priority

The study team assigned a recommendation for short-term or long-term priority for installation as well as a rank order to each recommended sign (see Table 6 and Table 7, columns one and two). These categories should inform park staff which routes or signs to prioritize. The implementation timeframe and rank is based on ease of installation, jurisdiction, distance from park, traffic volume, and visibility.

Signs with short-term priority (rank of 1-11) would be very helpful in directing motorists to the park. These signs are located along popular routes where there are no signs or an insufficient number of park signs. Signs with long-term priority (rank of 12-17) would improve existing signage to the park or potentially attract additional visitors traveling through the region. These signs have a long-term priority because they might be in a jurisdiction that is difficult for which to apply for new signs or the signs supplement existing signs.

Sign locations

There are three jurisdictions with which the park should focus signage investments and coordination efforts in order to improve visitor wayfinding to the park: the NY Thruway Authority, the city of Rome, and NYSDOT.

NY Thruway signs

If the park prefers to first concentrate on implementing new signs and wayfinding efforts along the NY Thruway, the park should first communicate their plans with the NY Thruway Authority. The following locations include new signs along the NY Thruway. The routes along NY Thruway are listed below with the route name, exit number, and new sign numbers, which correspond to Table 6 and Figure 33.

- *Route A: Syracuse/I-90/NY365 - Exit 33 (proposed signs 3-4 and 3-5; priorities 17 and 8)*
  There are two signs along this route under NY Thruway jurisdiction (see Figure 27). The study team recommends replacing these signs or adding two new signs along this route. If the signs are replaced, the park should work to replace the existing blue and white signs with brown and white signs to distinguish the signs from the background color and improve visibility. Alternatively, the park should work with the NY Thruway Authority to install new larger stand alone signs at or near this exit for westbound and eastbound traffic. These signs should be similar to the Erie Canalway NHC sign (Figure 32) with large font and the NPS arrowhead.

- *Route G: Herkimer/NY49 - Exit 31 (sign number 3-6; priority 9)*
  There are no signs along this route within the NY Thruway jurisdiction. The study team recommends adding one new sign along this route. Westbound motorists who research directions on a mapping internet site, such as Google Maps, are told to use exit 31 and follow NY 49 to the park. The park should work with the NY Thruway Authority to install sign 3-6 for westbound motorists at exit 31 that directs motorists to use exit 32.

- *Route I: Albany/NY69/I-90 - Exit 32 (signs 3-7 and 3-8; priorities 14 and 10)*
  There are two signs along this route (see Figure 27). The study team recommends the same options as for the signs along Route A above.

Rome signs

New signs within Rome would improve the visibility of the park to visitors, motorists passing through Rome, and the local community. The advantage of more signs in Rome is that these signs increase community awareness of the park and encourage visitation for motorists who are traveling near the park. These signs can improve wayfinding from other nearby historic sites, recreation attractions, or downtown Rome. The following locations include possible improvements to existing signs as well as additional signs on roads under NYSDOT and city of Rome jurisdiction. The routes are listed in order below with the route name and new sign numbers, which correspond to Table 7 and Figure 33 and Figure 34.

- *Route A: Syracuse/I-90/NY365 (sign 4-1; priority 1)*
  There is one sign along this route. The study team recommends adding an additional sign along this route since it is one of the most traveled routes into Rome from the NY Thruway. Sign 4-1 should be installed near the intersection with Black River Boulevard/NY26/NY46 to improve wayfinding at this intersection and along this route for northbound motorists. As shown in Table 6 and Table 7, the sign should include navigation directions and parking information.

- *Route B: Rome Sand Plains/Erie Boulevard West/NY46/NY49 (sign 4-2; priority 5)*
  There are four signs along this route. The study team recommends adding one new sign along this route. Sign 4-2 should be installed northwest of the intersection of Erie Boulevard West/NY46/NY49 and North James Street to improve wayfinding for motorists traveling southeast into Rome. The current brown and white historic site sign in the roadway median directs motorists into the left lane; however, it does not provide visitors with park information. A large brown and white NPS sign on the right side of the road should communicate park amenities, parking, and directions.

- *Route C: Taberg/NY69 (sign 4-2; priority 5)*
  There are four signs along this route. The study team recommends adding one new sign along this route that is the same sign described in Route B. As in route B, sign 4-2 for southeast bound motorists at the intersection of Erie Boulevard West/NY46/NY49 and North James Street would improve wayfinding along this route.

- *Route D: West Leyden/NY26 (sign 4-3; priority 7)*
  There are no signs along this route. The study team recommends adding one new sign along this route. Sign 4-3 at the intersection of Turin Street/NY26, North James Street, and East Bloomfield Street for southbound motorists would improve wayfinding to the park. The sign should be installed on the west side of Turin Street and should instruct motorists to veer right onto North James in order to reach the park, park amenities, and parking.

- *Route E: Boonville/NY46 (signs 3-1, 4-4, 4-5, and 4-6; priorities 15, 12, 13, and 4)*
  There are two signs along this route. The study team recommends adding four new signs along this route since they would improve wayfinding from the town of Boonville and Delta Lake State Park to the park. Sign 3-1 at the exit road leading out of Delta Lake State Park would inform southbound motorists of the distance and direction to the park. Closer to Rome, sign 4-4 at the intersection of Black River Boulevard/NY46 and East Chestnut Street would improve wayfinding

along this route. Sign 4-5, also at this intersection on East Chestnut Street, would improve wayfinding to northwest bound motorists on East Chestnut from Griffiss Air Force Base. Last, sign 4-6 at the intersection of Black River Road/NY46 and Floyd Avenue would improve wayfinding as southbound motorists approach the park. This sign should also include park amenities and parking information.

- *Route F: Holland Patent/Prospect/NY365 (signs 3-2 and 4-7; priorities 16 and 2)*
  There are no signs along this route. The study team recommends adding two new signs along this route since East Dominick Street is one of the main corridors leading into downtown Rome. Eastbound motorists coming from the Griffiss Air Force Base and Technology Park travel on Dominick Street into downtown Rome. Sign 3-2 at the intersection of Wright Drive/NY825 and East Dominick Street/NY365 would direct southbound motorists to East Dominick Street. In addition, sign 4-7 on East Dominick Street near the intersection with Black River Road/NY46 would improve wayfinding for westbound motorists along this route. This sign should also include park amenities and parking information.

- *Route G: Herkimer/NY49 (sign 4-8; priority 3)*
  There are two signs along this route. The study team recommends adding one new sign along this route near the existing sign on NY49/NY69/NY365 for northwest bound motorists. This sign should be similar to the Erie Canal NHC sign and include the NPS logo and mileage to the park.

- *Route H: Oriskany/NY69 (signs 4-1, 4-8, 4-9, and 3-3; priorities 1, 3, 6, and 11)*
  There are three signs along this route. The study team recommends adding four new signs along this route. Sign 4-1, described in Route A, and sign 4-8, described in Route G, would improve wayfinding for northwest bound motorists as they approach the park. Sign 3-3 at the exit of Oriskany Battlefield would improve wayfinding for motorists traveling west bound between the sites, as would sign 4-9 on Rome-Oriskany Road/NY 69 at NY233.

- *Route I: Albany/NY69/I-90 (sign 4-8; priority 3)*
  There are three signs along this route. The study team recommends adding one new sign along this route. Sign 4-8, described in Route G, would improve wayfinding for northwest bound motorists along this route.

Table 6

Proposed list of signs in the vicinity of Rome (Figure 33 / Map 3)

Source: Volpe Center

Rank (1 is top priority)	Short-term (ST), Long term (LT)	Route	Sign #	Road name	Intersection	Sign Text	Jurisdiction	Description
15	LT	E	1	NY46	Delta Lake State Park Road	right (6 miles)	NYSDOT	NPS sign with arrowhead, navigation arrow, and park distance
16	LT	F	2	Wright Drive	East Dominick Street	right (3 miles)	city of Rome/ NYSDOT	NPS sign with arrowhead, navigation arrow, and park distance
11	ST	H	3	Rome-Oriskany Road	Oriskany Battlefield road	right (6 miles)	NYSDOT	NPS sign with arrowhead, navigation arrow, and park distance
17	LT	A	4	I-90	Exit 33 westbound	Fort Stanwix National Monument, exit 33	Thruway Authority	NPS sign with arrowhead, navigation arrow, and park distance
8	ST	A,I	5	I-90	Exit 33 eastbound	Fort Stanwix National Monument, exit 33 or use exit 32	Thruway Authority	NPS sign with arrowhead, navigation arrow, and park distance
9	ST	G,I	6	I-90	Exit 31 westbound	Fort Stanwix National Monument, use exit 32	Thruway Authority	NPS sign with arrowhead, navigation arrow, and park distance
14	LT	I	7	I-90	Exit 32 westbound	Fort Stanwix National Monument, exit 32	Thruway Authority	NPS sign with arrowhead, navigation arrow, and park distance
10	ST	I	8	I-90	Exit 32 eastbound	Fort Stanwix National Monument, exit 32	Thruway Authority	NPS sign with arrowhead, navigation arrow, and park distance

Table 7

Proposed list of signs in Rome (Figure 34 / Map 4)

Source: Volpe Center

Rank (1 is top priority)	Short-term (ST), Long term (LT)	Route	Sign #	Road name	Intersection	Sign Text	Jurisdiction	Description
1	ST	A, F, G, H, I	1	Erie Boulevard	East Dominick Street	straight arrow, use right lane	city of Rome/ NYSDOT	NPS sign with arrowhead, navigation arrow, parking information and park distance
5	ST	B, C	2	Erie Boulevard	North James Street	straight arrow	city of Rome/ NYSDOT	NPS sign with arrowhead, navigation arrow, parking information and park distance
7	ST	D	3	Turin Street/NY26	North James Street and East Bloomfield	bear right (.5 miles)	city of Rome/ NYSDOT	NPS sign with arrowhead, navigation arrow, parking information, and park distance
12	LT	E	4	Black River Boulevard	East Chestnut Street	straight (2 miles)	city of Rome/ NYSDOT	NPS sign with arrowhead, navigation arrow, and park distance
13	LT	E	5	East Chestnut Street	Black River Boulevard	left (2 miles)	city of Rome/ NYSDOT	NPS sign with arrowhead, navigation arrow, and park distance
4	ST	E	6	Black River Boulevard	Floyd Ave	straight (.5 miles)	city of Rome/ NYSDOT	NPS sign with arrowhead, navigation arrow, parking information and park distance
2	ST	F	7	East Dominick Street	River Street	straight (1 mile)	city of Rome/ NYSDOT	NPS sign with arrowhead, navigation arrow, parking information, and park distance
3	ST	G, H, I	8	NY49	downtown Rome exit	bear right (1.5 miles)	NYSDOT	NPS sign with arrowhead, navigation arrow, and park distance
6	ST	H	9	Rome-Oriskany Road	NY 233	right (2 miles)	NYSDOT	NPS sign with arrowhead, navigation arrow, and park distance

Figure 33
Map 3 - Proposed signs
Source: Volpe Center

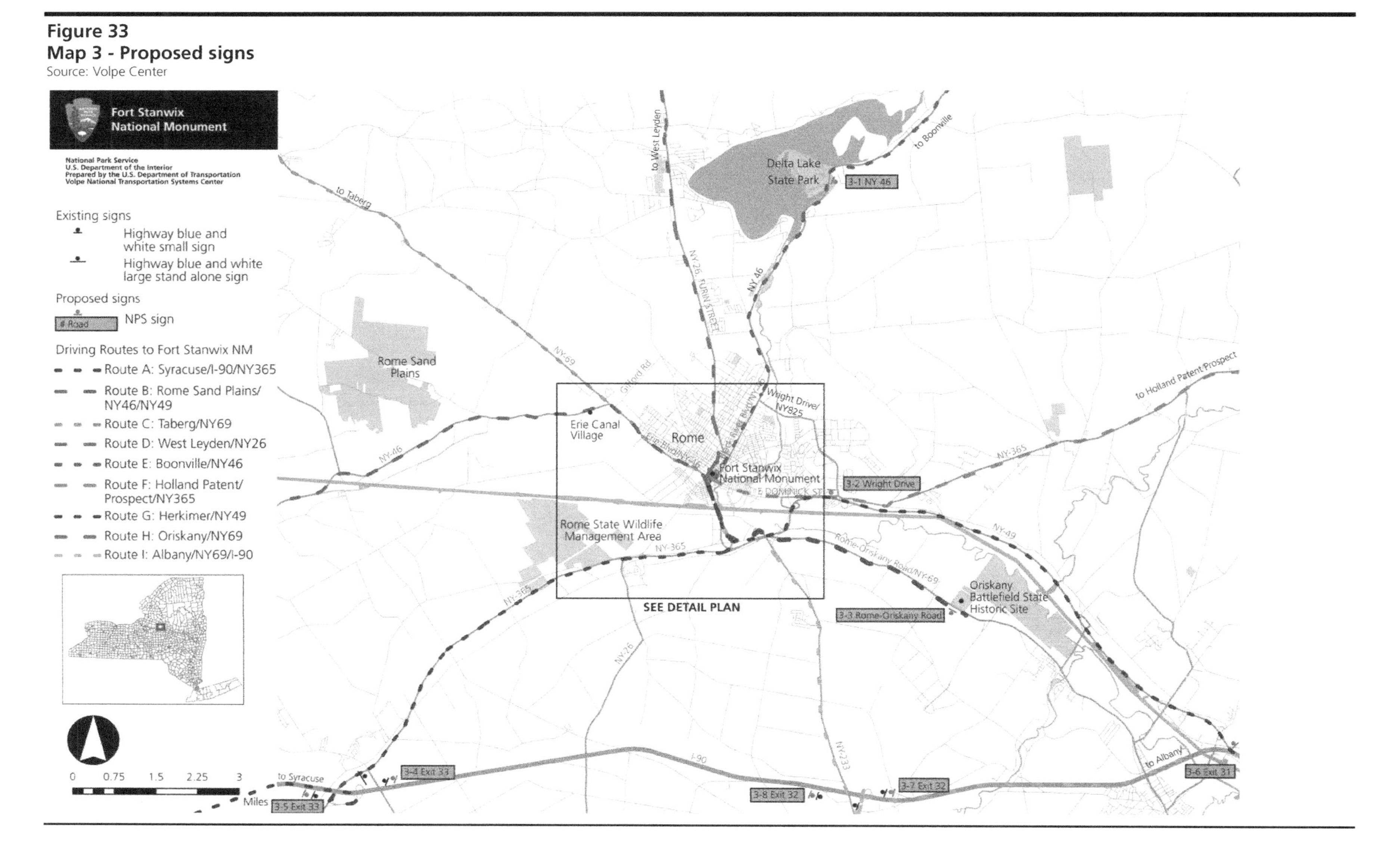

Figure 34
Map 4 - Proposed Rome signs
Source: Volpe Center

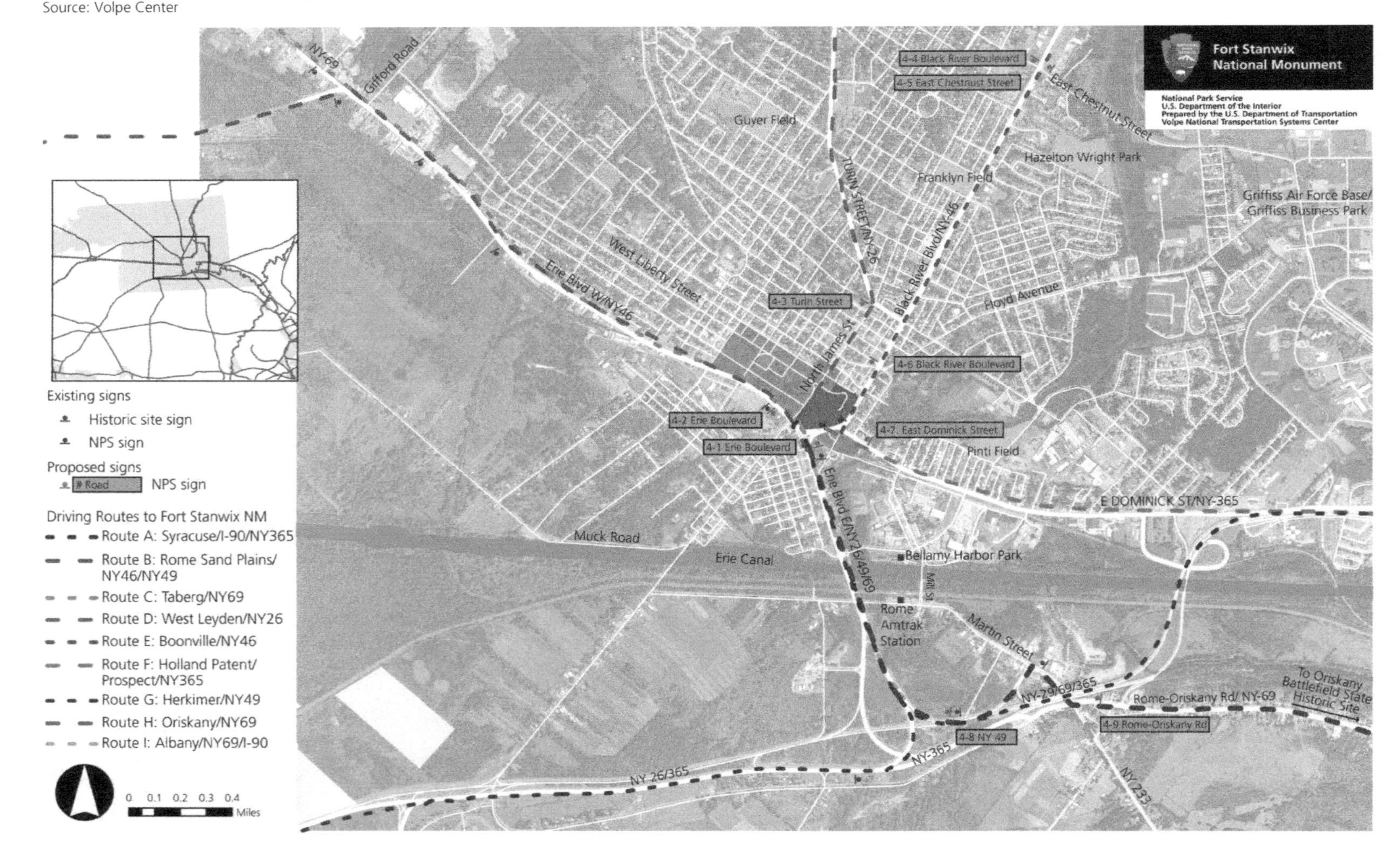

Sign investment analysis

There are several possible locations for new signs for the short- and long-term. The park can enhance wayfinding to the park if it invests in a couple of major routes to the park and directs visitors to these routes on the website, brochures, or when staff gives directions in person or over the phone. The following are four alternative strategies for investing in new signs.

- *Option A: The park improves signage from the NY Thruway at exit 32 to the park.* This option involves installing additional signs at exit 33 eastbound and exit 31 westbound to inform motorists to take exit 32. Exit 32 is the preferred exit from the NY Thruway because of the minimal turns a motorist has to make to access the park. Currently, internet directions guide motorists to take exit 31 when traveling westbound towards the park; therefore, under this option there would be signs east of exit 31 that direct visitors to take exit 32 to the park. Similarly, internet directions guide motorists to take exit 33 when traveling eastbound towards the park; therefore, under this option there would be signs west of exit 33 that direct visitors to take exit 32 to the park. This option directly improves Routes A, G, and I.
- *Option B: The park improves signage within the city of Rome.* This option involves increasing the frequency and visibility of signs along major routes to the park within the city of Rome. These routes include Erie Boulevard West/NY46/NY49, East Dominick Street/NY365, and Erie Boulevard/NY26/NY49/NY69. This option would improve local visibility and wayfinding to the park and improves parts of all Routes.
- *Option C: The park improves signage along NY Thruway exits.* This option involves installing new signs or improving signs at exits 31, 32, and 33 for eastbound and westbound traffic. This option directly improves Routes A, G, and I.
- *Option D: The park improves signage from local historic or recreation sites to the park.* This option involves installing new signs from Oriskany Battlefield, Erie Canal Village, the Erie Canal, and Delta Lake State Park. This option directly improves Routes B, E, and H.

Process

The NPS is responsible for signs within the park boundary. Depending on their location outside of the park boundaries, coordination with the city of Rome, NYSDOT, and the NY Thruway Authority is necessary in order to apply for and install new vehicular wayfinding signs.

NYSDOT is responsible for all routes with a prefix of NY on the roadway name. The park should contact the Region 2 Real Estate Officer in Utica in order to apply for new signs.[1] An example of this sign is in Figure 35. In 2010, the contact is Lisa Rowlands (315) 793 2405 or Deborah Windecker, Regional Real Estate Officer, (315)793-2412. For signs along NYSDOT roads that are within the city of Rome, coordination with the city of Rome should also occur.

The NY Thruway Authority is responsible for signs along the NY Thruway. In order to apply for new signs or changes to existing signs along the NY Thruway, the NPS should contact the NY Thruway office, Syracuse Division in Liverpool. The park can contact the Highway Department directly at (315)438-2324 to discuss the sign options.[2]

---

[1] Phone conversation with Lisa Rowlands (315) 793 2405. July 12, 2010. Additional contact and application information: https://www.nysdot.gov/regional-offices/region2/contacts.
[2] Phone conversation with Patti Traver-Fleming (315)438-2324. July 12, 2010. Additional contact information: http://www.thruway.ny.gov/about/addresses.html#syracuse.

Figure 35
Historic site sign
Source: NYSDOT

SIGN DRAWING SD-G16	Brown Background	White Legend
Historic Site Signs (NYM9-1, NYM9-2, NYM9-3, NYM9-4)		

NYM9-1

NYM9-2

NYM9-3

NYM9-4

	Sign	Size	Border	Line 1	Line 2	Line 3
C	NYM9-1	30" x 24"	Varies	4"-D	2"-F	
	NYM9-1	45" x 36"	Varies	6"-D	3"-F	
C	NYM9-2	30" x 24"	Varies	4"-D	4"-D	2"-F
	NYM9-2	45" x 36"	Varies	6"-D	6"-D	3"-F
C	NYM9-3	30" x 24"	Varies	4"-D	2"-D	
	NYM9-3	45" x 36"	Varies	6"-D	3"-D	
C	NYM9-4	30" x 24"	Varies	4"-D	4"-D	2"-D
	NYM9-4	45" x 36"	Varies	6"-D	6"-D	3"-D

Sign type

There are three existing types of signs for the park that are intended for motorists. These include blue and white signs along the NY Thruway, brown and white historic site signs, and brown and white signs on the park property (Figure 24). For all new signs, the park should consider using a consistent design and brown and white color scheme with the NPS arrowhead. Larger signs with larger text are appropriate for roadways with higher average speeds, while smaller-text signs are appropriate for lower speed roadways.

Cost

Cost of permit fees, signs, and sign installation vary depending on the jurisdiction. For the NYSDOT signs, there is a permit cost per year depending on the type of sign. A conversation with NYSDOT on specific signs will provide NYSDOT with the information they need to determine the permit cost.

New NPS signs should comply with the Manual of Uniform Traffic Control Devices (MUTCD). The MUTCD is published by the Federal Highway Administration (FHWA) and defines the standards used by road managers for the installation and maintenance of traffic control devices on all public roads and highways.[1] While sign costs vary, there are pricings available for similar signs.[2] A typical steel post sells for $42, a stand is $15, and the sign is $40-$50. The total price for a new sign and post is approximately $97-$107 per sign. The exact cost will depend on the size, installation, and location of the sign. NYSDOT typically purchases signs through *Corcraft* or *Corrections Craft,* which sells to government entities and non-profit organizations. Through *Corcraft*, the 24" x 30" "Historic Site" signs are $28.55 each (Figure 35). The 15" x 21" signs with an arrow, which can be added underneath the "Historic Site" signs, are $13.10 each. Other agencies may have specific sign types, size requirements, and installation requirements that are different from NYSDOT sign regulations. Jurisdictions should be contacted individually for specific cost and sign information.

Partnerships

The park is within the Erie Canalway NHC. To attract visitors in this corridor who are already interested in historic and recreation sites, the park should coordinate with the Erie Canalway NHC and other sites within the corridor to install signs at nearby and related historic and recreation sites and to coordinate sign applications with the NYSDOT and NY Thruway Authority. This coordination would improve wayfinding throughout the corridor so that visitors can better navigate between sites throughout the region.

Conclusion

This report identifies and prioritizes 17 new signs that would benefit motorists traveling to the park. The implementation timeframe and rank is based on ease of installation, jurisdiction, distance from park, traffic volume, and visibility. Signs with a shorter-term priority (rank of 1-11) are necessary to direct motorists traveling to the park. Signs with a longer-term priority (rank of 12-17) will improve existing signs or attract additional visitors traveling through the region.

---

[1] U.S. Department of Transportation Federal Highway Administration. Manual on Uniform Traffic Control Devices. http://mutcd.fhwa.dot.gov/.

[2] Rice Signs: Your Leader in Transportation Safety. Traffic signs. http://www.ricesigns.com/?gclid=CMqE5-3Rn5UCFQNHFQodlW6kjw.

Parking

Parking at Fort Stanwix has been a problem for several years; however, the issue is not due to a lack of nearby, available parking spaces. Rather, the problem stems from the fact that all parking options around the park are owned and managed by entities other than NPS. As a result, the park has little control over parking location, fees, accessibility, signage, facilities, management, and maintenance. Furthermore, the park is unable to assure visitors that their cars will not be ticketed or towed, that their visits will not be cut short due to time restrictions, or that their vehicles are secure.

Currently Fort Stanwix has no preferred parking alternative and cannot afford to invest time, money, or resources into an unproven parking solution. The purpose of this section of the report is to (1) identify the park's current parking arrangements and actual visitor parking practices, (2) assess the park's future parking needs, (3) identify parking alternatives, (4) assess the strengths and weaknesses of each parking alternative, and (5) recommend alternatives for moving forward with a long-term goal of finding a permanent parking solution for Fort Stanwix.

Fort Stanwix current parking arrangements and visitor parking practices

Staff at Fort Stanwix are constantly faced with the question of where they can tell visitors to park. Because the park does not control any of its parking alternatives, it is difficult to provide clear direction to visitors.

In past years, the park was able to direct all autos, recreational vehicles (RVs), and buses to the North James Street Parking Lot, on the corner of North James Street and West Dominick Street, across North James Street from the Willett Center (Figure 36). Recently, Rome Savings Bank, the owner of the North James Street Parking Lot, expressed concerns regarding liability associated with park visitors parking in their lot. The bank asked Fort Stanwix to direct its visitors to park elsewhere and no longer clears its lot during winter months.

Figure 36
North James Street Parking Lot, looking east toward the park's Visitor Center (left) and looking southwest across the intersection of North James and West Dominick Streets (right)
Source: Volpe Center

Currently, the park's website directs visitors to park in the 525-space Fort Stanwix Parking Garage on the corner of North James Street and Liberty Avenue (Figure 37). According to reports from the city's parking department, the garage averages only two to three non-permit entrances per day[1]. This level suggests that

[1] Reported from city of Rome Director of Community & Economic Development Director Diane Shoemaker, May 27, 2010.

most park visitors who arrive by car are *not* parking in the garage; field observations by the study team confirm this practice. Fort Stanwix staff indicate that the vast majority of park visitors still usually park in the North James Street Parking Lot. Again, field observations support this characterization. A handful of visitors also park in designated street parking spaces along West Dominick Street, East Park Street, and Church Street. Visitors also occasionally park in the small off-street parking area behind the administration office building and behind St. Peter's Roman Catholic Church on the corner of North James and East Park streets.

Figure 37
Fort Stanwix parking garage
Source: Volpe Center

Fort Stanwix parking needs assessment

The next step of the analysis is to estimate the number of parking spaces the park actually needs to accommodate visitors at the busiest time of year. Since most non-bus visitation occurs on the weekends, the parking needs assessment (Table 8) focuses on weekend days from 2007 and 2009. The assessment allows the park to compare current and future estimated needs with the number of available spaces in existing facilities surrounding the park.

In order to determine a realistic needs assessment, special event visitation is not a factor in determining daily parking needs. The reason for this exclusion is that a viable parking needs assessment must focus on the needs of the park for the majority of the days that the park is open, rather than for a few select occasions that occur only once a year. For a few major events at the park, visitation is many times higher than the average day. For these events, parking efforts will *always* require special coordination between the park, the city, and surrounding landowners to ensure safe access and efficient traffic flow.

The only hard data available for the needs assessment include total visitation per month, visitation for special events per month, and number of buses per month. Because of a lack of detailed daily visitation data, the analysis uses a number of assumptions based on information provided by the park and other factors commonly used in determining parking needs. The assessment can be easily updated to reflect changes in these assumptions.

Table 8
Fort Stanwix parking needs assessment
Source: Volpe Center

Month	Average monthly visitation excluding special events and bus visitors	Weekend visitors excluding special events and bus visitors	Weekday visitors excluding special events and bus visitors	Weekend daily visitors excluding special events and bus visitors	Weekday daily visitors excluding special events and bus visitors	Passengers per vehicle	Vehicles per weekend day	Parking Space Turnover	Weekend Parking Spaces Needed
January	420	210	210	26	10	2.5	10	2	5
February	317	159	159	20	7	2.5	8	2	4
March	589	294	294	37	13	2.5	15	2	7
April	2,816	1,408	1,408	176	64	2.5	70	2	35
May	3,037	1,519	1,519	190	69	2.5	76	2	38
June	4,371	2,186	2,186	273	99	2.5	109	2	55
July	9,187	4,594	4,594	574	209	2.5	230	2	115
August	7,194	3,597	3,597	450	164	2.5	180	2	90
September	3,942	1,971	1,971	246	90	2.5	99	2	49
October	2,868	1,434	1,434	179	65	2.5	72	2	36
November	1,188	594	594	74	27	2.5	30	2	15
December	339	170	170	21	8	2.5	8	2	4

Parking needs assumptions

The parking needs assessment assumes that:

- All visitors arrive by car or bus. The assessment does not account for the number of visitors who walk or bike to the park as there is no data to provide an accurate pedestrian/bicycle count.
- Each bus carries an average of 40 passengers.
- Weekends experience higher non-bus visitation than weekdays, with 50 percent of non-bus visits to Fort Stanwix occurring during the week (Monday-Friday), throughout the year, and fifty percent of non-bus visits to Fort Stanwix occurring on weekends (Saturdays and Sundays). In other words, half of the non-bus visits are condensed into eight days of a typical month, whereas the other half is spread over twenty-two days.
- Each private automobile carries an average of 2.5 passengers.
- The average length of visit to Fort Stanwix is half of one day. This assumption means each parking space can be used twice each day.

Parking needs results

The parking needs assessment shows that weekends in July are the busiest time of year for non-bus visitors. Fort Stanwix requires approximately 115 parking spaces to accommodate every vehicle on an average July weekend day. This is likely a conservative estimate since at least some visitors will be arriving at the park by bicycle or on foot. By a significant margin, July is the busiest month, potentially requiring twenty-five more spaces on a weekend day than in August, the next busiest month (Figure 38).

Figure 38
Number of parking spaces needed at Fort Stanwix on an average weekend day, by month
Source: Volpe Center

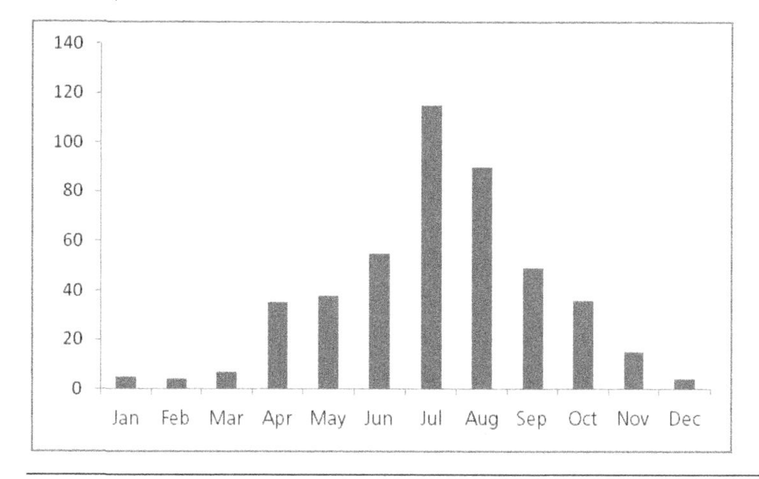

Based on total annual visitation trends over the past three years, it is projected that park visitation will continue to grow in the near future. If this proves to be true, the number of cars will grow, and the park's parking space needs will increase accordingly. Assuming a proportional relationship between visitation and parking needs, Table 9 shows the number of spaces needed for a weekend day in July for future years for a variety of potential growth rates. Most importantly, this table shows that by the time visitation doubles, the maximum number of parking spaces that would be needed is approximately 230.

Table 9
Future parking needs
Source: Volpe Center

Month	Average weekend day parking spaces needed 2007-2009	Spaces needed if park visitation grows by: 1%	3%	5%	10%	25%	100%
January	5	6	6	6	6	7	11
February	4	4	5	5	5	5	8
March	7	8	8	8	9	10	15
April	35	36	37	37	39	44	71
May	38	39	40	40	42	48	76
June	55	56	57	58	61	69	110
July	115	116	119	121	127	144	230
August	90	91	93	95	99	113	180
September	49	50	51	52	55	62	99
October	36	37	37	38	40	45	72
November	15	15	16	16	17	19	30
December	4	5	5	5	5	6	9

Existing parking alternatives

Visitors to Fort Stanwix may park in nearby private parking lots, the city-owned Fort Stanwix Parking Garage, or in a number of designated street-parking spaces. Based on current parking trends and potential future parking opportunities, all relevant parking locations and characteristics are listed below, and the location of each is shown in Figure 39.

1) Fort Stanwix Parking Garage
   - Owned and managed by the city of Rome
   - 522 spaces
   - Fee to park: $1.00 will pay for the first hour, and subsequent hours cost $0.50 each
   - Closed on weekends unless opened by city legislation
2) North James Street Parking Lot
   - Owned and managed by Rome Savings Bank[1]
   - Approximately 85 spaces
   - No fee to park; parking is restricted to bank-related uses only
3) West Dominick Street diagonal on-street parking
   - Owned and managed by the city of Rome
   - Approximately 60 spaces
   - No fee to park; two-hour time limit
4) Church Street (diagonal) and East Park Street (parallel) on-street parking
   - Owned and managed by the city of Rome
   - Approximately 26 spaces
   - No fee to park; 90 minute time limit

[1] Before publication of this report, Rome Savings Bank was acquired by Berkshire Bank. The parking restriction remains the same.

5) Erie Boulevard access road parking area
   - Owned and managed by the city of Rome
   - Approximately 21 spaces

Figure 39
Parking alternative locations
Source: Volpe Center

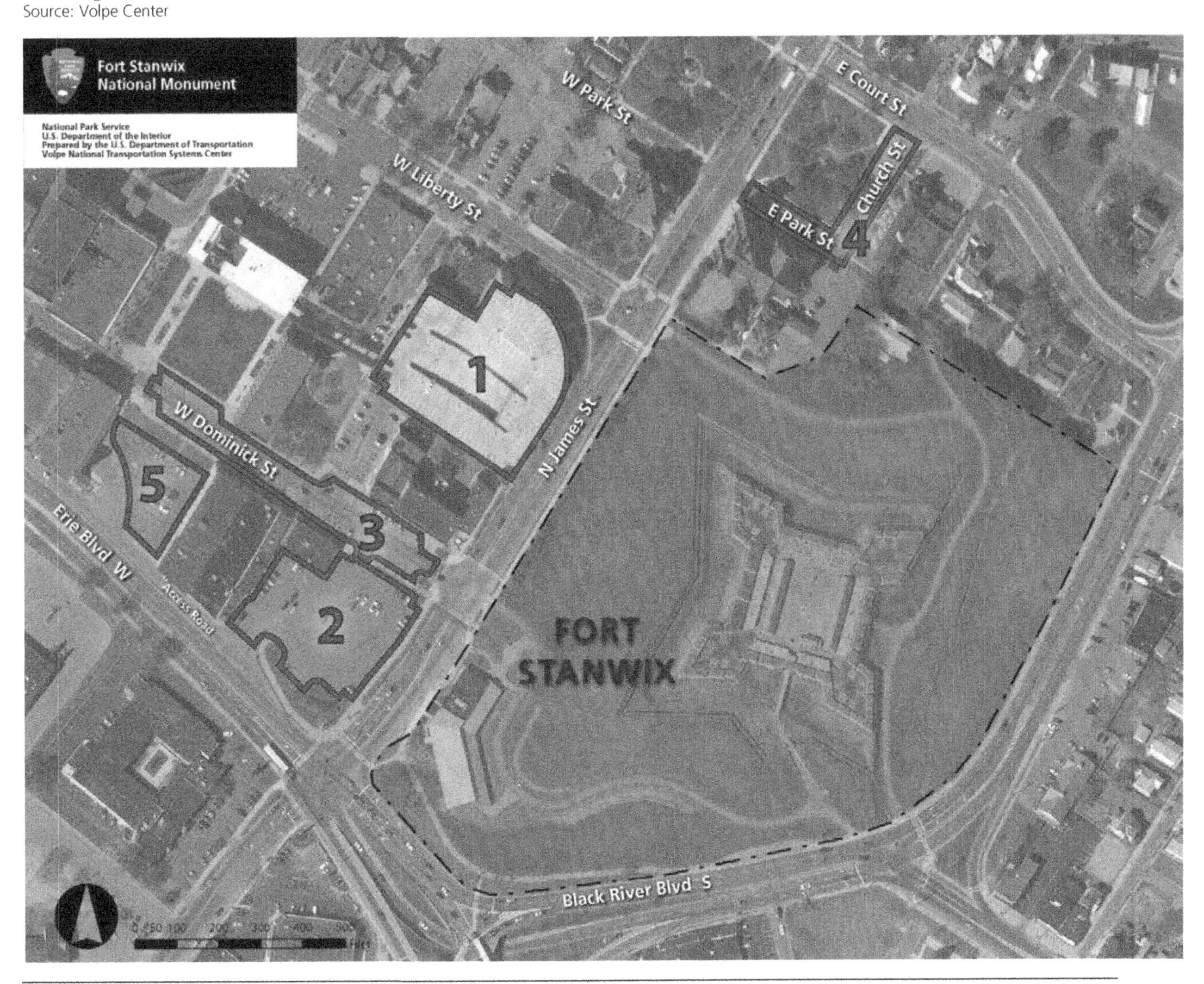

An occupancy analysis provides an accurate assessment of parking opportunities for each alternative. The analysis intends to show where parking is actually available at a given time of day on a given day of the week, informing all parties of the true situation on the ground.

For the purposes of this study, the occupancy analysis shows which parking areas have enough vacant spaces to support the demand for parking generated by Fort Stanwix. Table 10 shows general specifications and the weekday occupancy rates of each parking area and the expected weekend parking availability for the Fort Stanwix Parking Garage. The study team was not able to conduct weekend space counts for the other parking areas under consideration.

Table 10
Specifications and available spaces for alternative parking areas[1]
Source: Volpe Center

Map I.D.	1	2	3	4	5
Site Name	**Fort Stanwix Garage**	**North James Street Parking Lot**	**West Dominick Street Diagonal On-street Parking**	**Church St/ East Park St On-street Parking**	**Erie Boulevard Access Road Lot**
Location	North James Street & Liberty Avenue	North James Street & West Dominick Street	West Dominick Street corridor, west of North James Street	Church Street, south of E Court Street	Off Erie Boulevard Access Road
Owner/ manager	city of Rome	Rome Savings Bank	city of Rome	city of Rome	city of Rome
Cost to Park	$1.00 per hour weekdays; free on weekends if open	Free	Free (2 hour limit)	Free (2 hour limit)	Free
Number of Parking Spaces	522	85	60	26	21
Estimated Number of Available Spaces (weekday AM)	250[2]	52	12	18	2
Estimated Number of Available Spaces (weekday PM)	250	60	24	19	3
Estimated Number of Available Spaces (weekend AM)	250+	N/A	N/A	N/A	N/A
Estimated Number of Available Spaces (weekend PM)	250+	N/A	N/A	N/A	N/A

Upon examining the existing parking alternatives, it is clear that the Fort Stanwix Parking Garage has the greatest capacity and the most available spaces on weekdays. The North James Street lot appears to have the second greatest capacity on weekdays.

Also, since the Fort Stanwix Parking Garage is primarily used by downtown workers during regular business hours (and is generally closed on weekends), it can be assumed that many more than 250 spaces are available on Saturdays and Sundays. On an average weekend day in July, the garage can satisfy 100 percent of the park's parking needs. Furthermore, if the park's visitation were to *double*, the parking garage would still be able to accommodate 100 percent of the Fort's parking needs.

[1] The study team was not able to count weekend occupancy rates for the parking areas in question. Complete instructions were provided to the park should they be able to complete weekend parking counts with an available volunteer or staff person in the future. While the data will help to provide a more complete assessment of occupancy for each parking alternative, it would not likely affect our ultimate recommendations.

[2] Reported from city of Rome Director of Community & Economic Development Director Diane Shoemaker, May 27, 2010.

Parking challenges

Each of the potential parking areas discussed above has a number of challenges that prevent it from being the ideal parking choice for Fort Stanwix.

1) Fort Stanwix Parking Garage
   - By law, the city does not open the parking garage on weekends. In order to open the garage on weekends, the Rome City Council must pass a formal resolution. In the past, the city has successfully passed legislation allowing full accessibility to the garage at no charge during summer weekends.
   - Even when legislation has directed the garage to be opened on weekends, the garage has failed to be open on a number of occasions.
   - Parking garage height restrictions are unable to accommodate RVs and buses.
   - There is a fee to park in the garage. At the present time, $1.00 pays for the first hour, and subsequent hours cost $0.50 each.
   - Lower floors of the garage are usually fully occupied by monthly permit holders. Fort visitors have to park on the upper levels during the week.
   - Visitors have to cross North James Street to access the park. Despite recent improvements, the crossing is busy and potentially dangerous.

2) North James Street Parking Lot
   - Rome Savings Bank, the owner of the North James Street Parking Lot, has requested that its parking area no longer be used by park visitors due to future liability concerns.
   - Snow is no longer removed from the majority of the parking area during the winter months.
   - Visitors have to cross North James Street to access the park.

3) West Dominick Street diagonal on-street parking
   - The on-street parking spaces on West Dominick Street are limited to two hours.
   - The two-hour time restriction may cause visitors to shorten their visit to the park and/or limit a visitor's ability to patronize other shops/restaurants/businesses in the vicinity.
   - Visitors have to cross North James Street to access the park.

4) Church Street (diagonal) and East Park Street (parallel) on-street parking
   - The on-street parking spaces on Church Street and East Park Street are limited to 90 minutes.
   - The 90-minute time restriction may cause visitors to shorten their visit to the park and/or limit a visitor's ability to patronize other shops/restaurants/businesses in the vicinity.
   - These parking spaces are located farthest from the Visitor Center, but the walk does not require any major street crossings.

5) Erie Boulevard access road parking area
   - Ingress and egress from the parking area are not intuitive and, due to roadway configurations, the lot is difficult to access from the park pick-up and drop-off area.
   - The area is highly occupied during weekdays, with only a few available spaces for cars.

Parking recommendations

The study team believes that Fort Stanwix should not consider development of any new parking. The city documented that parking space is abundant in the area surrounding the park[1], and preliminary occupancy analyses conducted by the study team suggest that spaces are always available. Given the existing unused parking space near the park, the study team believes the park should aim to satisfy long-term parking needs through an official agreement with the city of Rome and any other potential partners and stakeholders. Once stakeholders agree upon what lots park visitors should use, signage should be installed to direct visitors to the appropriate locations. Stakeholders' websites as well as visitor center staff should be sure to direct visitors to these locations as well.

Parking for cars

The Fort Stanwix Parking Garage is one of the city's most important parking facilities, and it may play an important role as the city introduces new visitors to a variety of downtown destinations. Unfortunately for the park (and potentially any other recreational destinations), the garage is not generally open on weekends (some of the park's busiest days), and the process for opening the garage during summer weekends is convoluted. However, in past years the garage has successfully been opened on weekends to provide free parking to anyone wishing to be downtown during the summer months, either to visit the park or for any other reason. In discussions with the city, it is clear that despite the difficulties of getting the garage opened, no negative consequences resulted from the garage being opened on the weekends. As long as that remains the case, subsequent requests by the park to open the garage on the weekends will likely be honored.

The study team recommends that Fort Stanwix should work with the city to investigate any opportunities or agreements that could ensure that the Fort Stanwix garage is available for parking seven days a week, year-round. Though it may require a significant effort in order to change the city's current garage policies, the result could produce an impact that positively affects all parties.

Opportunities for the park may include:

- The ability to direct all autos to a single site for parking.
- The ability to develop (a) vehicular signage that directs autos to parking, and (b) pedestrian signage that directs visitors from the parking area to the Visitor Center.
- The potential to develop a long-standing agreement with the city that is not influenced by private landholders.
- Covered parking during winter months.

Opportunities for the city may include:

- Parking revenue during the week (and weekend, if the garage is staffed).
- Less demand for surface parking, perhaps freeing land used for parking that may have redevelopment potential.
- The potential to capitalize on increased foot-traffic around the intersection of Dominick and James.

If the park wishes to provide free parking to visitors, opportunities may also exist for developing a pre-paid ticket validation program, whereby NPS purchases a number of validation stickers based on predicted need and average length of visit. Stickers may then be distributed to visitors who have parked in the garage.

On-street parking would be a good alternative or supplement to an established, paid-parking facility. On-street parking is free, but the time is limited to two hours. Space is available throughout the day on West Dominick Street, East Park Street, and Church Street for those visitors who do not

[1] City of Rome Urban Design Plan, 2004.

wish to pay for parking. As a way to provide park visitors parking on the same side of the block as the visitor center and the park, the park could work with the city to consider changing the time limit from two to four hours along East Park Street and Church Street.

Parking for buses and RVs

If the Fort Stanwix garage is the park's primary parking facility for standard automobiles, then locating suitable parking for buses and RVs is still a challenge. Similar to many autos, some buses and RVs currently use the North James Parking Lot.

The most promising alternative is the city-owned Erie Boulevard access road lot, situated between West Erie Boulevard and West Dominick Street (Figure 40). The lot is accessible only from the Erie Boulevard access road. Due to the roadway configuration between the park and this parking area, a bus or RV would have to travel around a number of city blocks in order to use this lot. Once the vehicle is parked, however, the lot offers access by stairway to West Dominic Street, approximately 150 yards from the intersection with North James Street.

Another concern with the Erie Boulevard access road lot is that it appears to be one of the more heavily occupied parking lots in the area. Depending on future use, the lot may not have room to accommodate large vehicles, particularly during a busy day during the school year when buses are most frequent.

Figure 40
Erie Boulevard access road lot: potential bus/RV parking between Erie Boulevard and Dominick Street
Source: Volpe Center

Conclusions

Because Fort Stanwix does not possess its own parking area, the question of where to park is an issue. However, numerous public and private options for parking do currently exist, and all are within easy walking distance of the park grounds and the Willett Center.

Given the park's location in the heart of Rome, it is essential that the park work closely with the city and other downtown stakeholders to ensure all decisions support downtown development initiatives. As one of Rome's most important tourist attractions, Fort Stanwix attracts visitors who are likely to spend money at businesses and other attractions in the area. With a clear parking policy in place, the park will be able to work with the city to ensure all parties – the city, the park, and local businesses – are able to benefit from the presence of visitors in downtown Rome.

Pedestrian access and safety

Pedestrian access and safety improvements in and around Fort Stanwix National Monument are necessary to improve visitor safety and circulation, as well as to enhance the overall experience of visiting the park and surrounding areas of downtown Rome. By improving pedestrian-oriented infrastructure, particularly at major intersections, park entrances, and along sidewalks and pathways, the park will improve the experience of its visitors and promote safe pedestrian access to and from adjacent streets and neighborhoods.

Purpose

This section of the report focuses on (a) identifying existing pedestrian facilities including sidewalks, crosswalks, intersections, and pedestrian amenities, (b) analyzing existing use of pedestrian facilities, and (c) developing prioritized recommendations.

Existing pedestrian network

Two major paths cross park property traveling between Black River Boulevard and North James Street (Figure 41). One path passes south of the actual park, connecting East Dominick Street to West Dominick Street via the visitor center. The other path runs north of the park, connecting East Dominick Street with West Liberty Street. Several short spurs and connectors complete the pedestrian network within the park boundary. All paths are highlighted in Figure 42.

Figure 41
Fort Stanwix Paths and Signage
Source: Volpe Center

An extensive network of sidewalks and crosswalks surround the park along North James Street, Erie Boulevard East, and Black River Boulevard (Figure 42 and Figure 43).

Figure 42
Existing pedestrian facilities
Source: Volpe Center

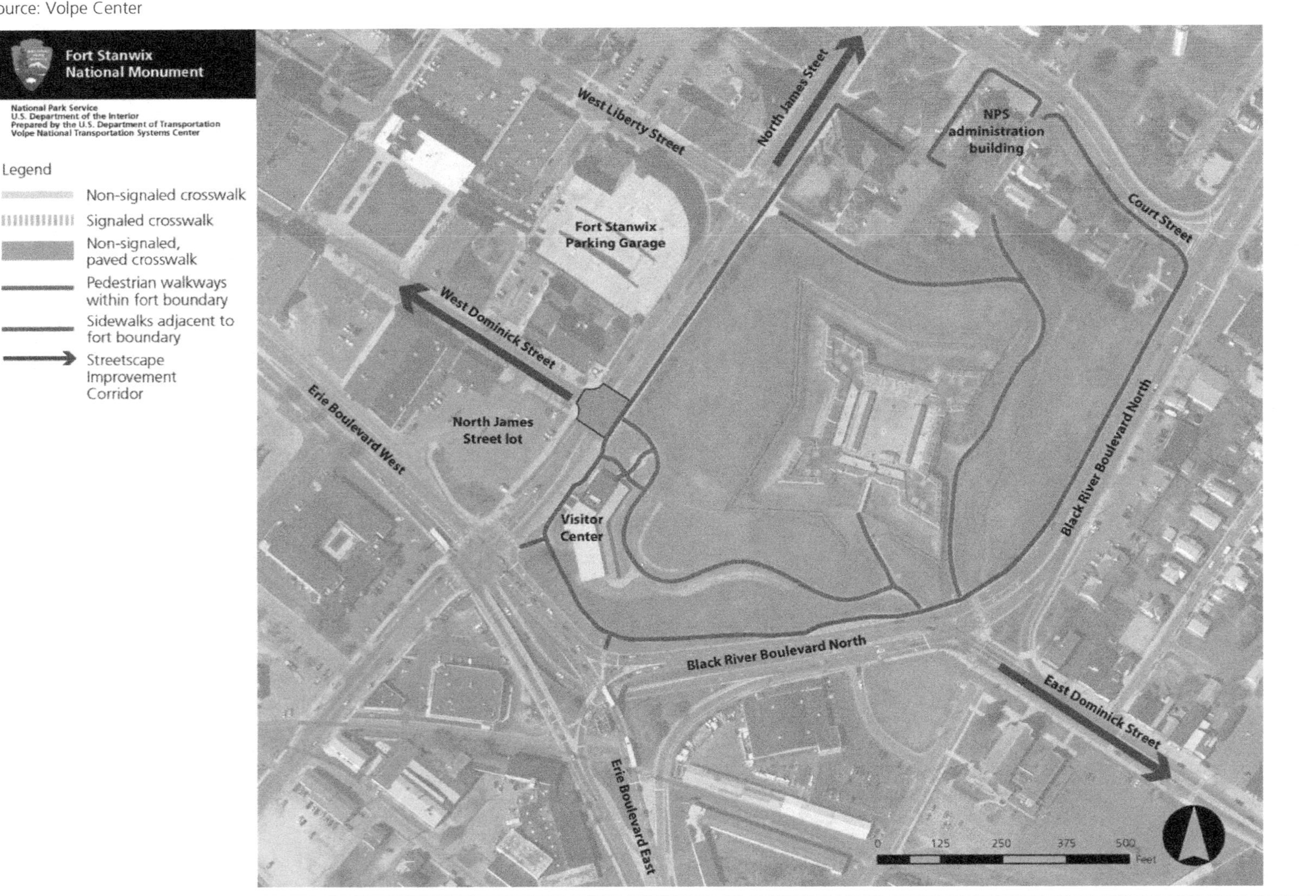

Figure 43
Sidewalks along North James Street (left) and Black River Boulevard (right)
Source: Volpe Center

Four signalized intersections adjacent to the park are outfitted with the following facilities:

- Three crosswalks and pedestrian signals at the intersection of North James Street and West Liberty Street. Pedestrians cross this intersection when they are walking between the park and the businesses along West Liberty Street or between the park and the Tomb of the Unknown Soldier at North James Street and West Liberty Street (Figure 44, left).
- Three crosswalks at the intersection of North James Street and Erie Boulevard. There are restaurants, a drugstore, and a hotel at this intersection across from the park.
- Three crosswalks at the intersection of Black River Boulevard and Erie Boulevard. These crosswalks pass over NY 46 and across Erie Boulevard East.
- Two crosswalks at the intersection of Erie Boulevard North and East Dominick Street. In addition, there is one non-signaled crosswalk on East Dominick Street (Figure 44, right).

Figure 44
Signaled crosswalks at North James St and West Liberty St (left) and Black River Blvd and Erie Blvd (right)
Source: Volpe Center

The intersection of North James Street and West Dominick Street includes a painted and textured crosswalk, a removable pedestrian-crossing sign in the middle of the crosswalk, and pedestrian-crossing signs on North James Street for motorists traveling in both directions (Figure 45).

Figure 45
North James Street and West Dominick Street textured crosswalk
Source: Volpe Center

Adjacent pedestrian districts

Fort Stanwix is situated at the confluence of three important pedestrian-friendly corridors within Rome. Each corridor has been targeted by the city for streetscape improvement, and Fort Stanwix is perfectly situated to serve as a central pedestrian hub of the city. The three corridors – West Dominick Street, East Dominick Street, and North James Street – are discussed on page 14 of the Existing Conditions Update for this project and are highlighted in Figure 42.

Pedestrian analysis

To assess existing pedestrian circulation, the study team observed pedestrian behavior at three locations around the perimeter of the park for three twenty-minute periods each. Shown in Figure 46, the three locations were:

- Site 1: Visitor center entrance near the intersection of North James Street and West Dominick Street
- Site2 : Intersection of Black River Boulevard and East Dominick Street
- Site 3: Intersection of North James Street and West Liberty Street

Observations took place on a weekday in June during the morning and early afternoon. The study team selected the three sites because of their proximity to intersections and park entrances.

Figure 46
North James Street and West Dominick Street Crosswalk
Source: Volpe Center

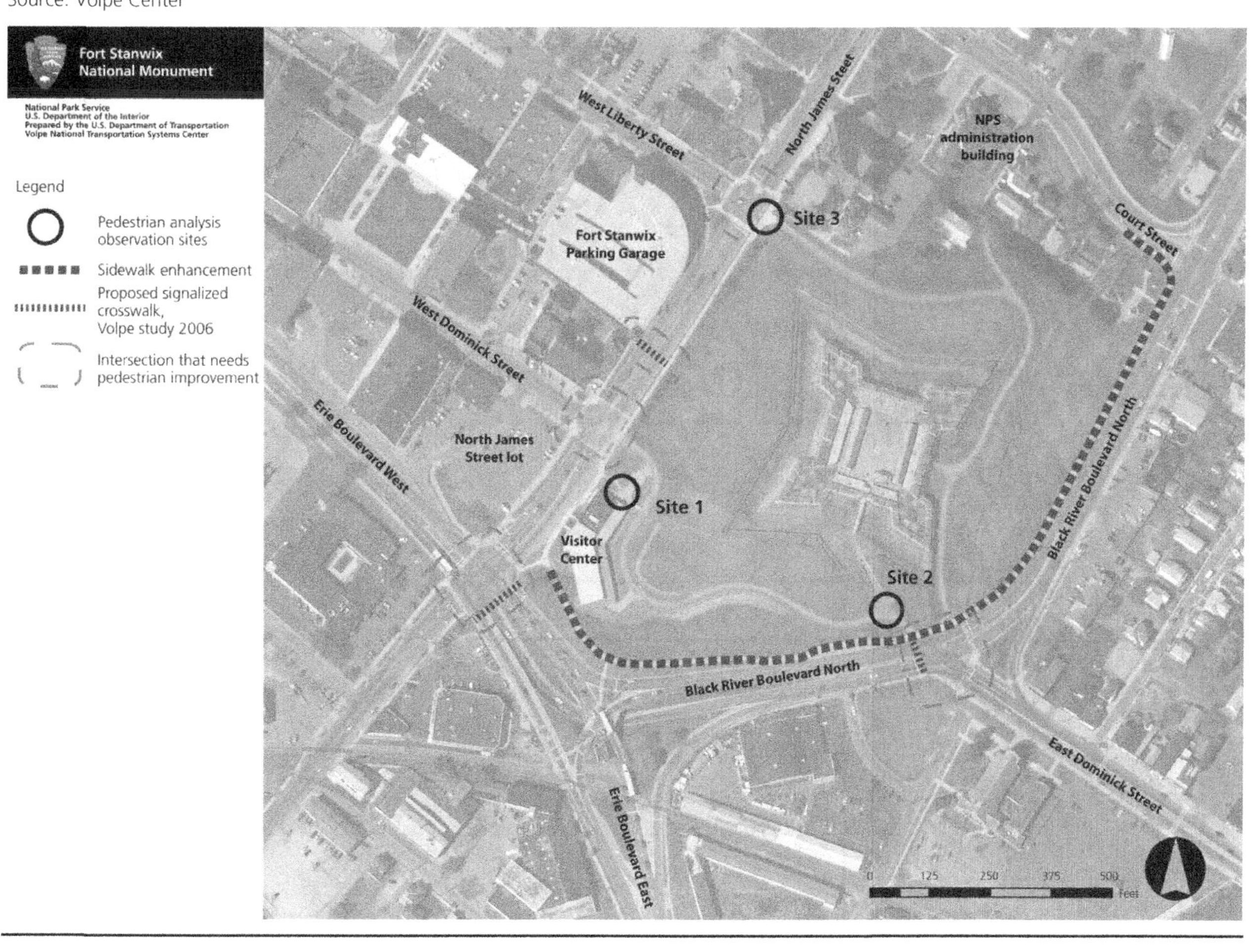

Non-motorized travel in and around the park consists primarily of pedestrians and bicyclists. Some are specifically traveling to the park, but the study team observed frequent use of the pathways by pedestrians and bicyclists passing through the premises, particularly en route between West Dominick Street and East Dominick Street and vice versa. Table 11, Table 12, and Table 13 illustrate pedestrian and bicycle activity in and around the park during the study team's three periods of observation at the three sites.

Table 11
Pedestrian and bicycle counts: Visitor Center Entrance (Site 1)
Source: Volpe Center

time:	intersection	Visitor Center / Fort ped	Visitor Center / Fort bike	Visitor Center / Fort total	Through site ped	Through site bike	Through site total
9:10-9:30	North James and West Dominick	1			2	2	
	Black River Boulevard and East Dominick St			1			5
	Did not cross a street					1	
10:10-10:30	North James and West Dominick	4			2		
	Black River Boulevard and East Dominick St			49			3
	Did not cross a street	45				1	
1:40-2:00	North James and West Dominick	4			1		
	Black River Boulevard and East Dominick St			8			3
	Did not cross a street	4			1	1	
total				58			11

Table 12
Pedestrian and bicycle counts: Fort Entrance near East Dominick Street (Site 2)
Source: Volpe Center

time:	intersection	Visitor Center / Fort ped	Visitor Center / Fort bike	Visitor Center / Fort total	Through site ped	Through site bike	Through site total
9:30-9:50	North James and West Dominick						
	Black River Boulevard and East Dominick St	3		3			1
	Did not cross a street					1	
10:30-10:50	North James and West Dominick						
	Black River Boulevard and East Dominick St			1		1	3
	Did not cross a street	1			2		
1:20-1:40	North James and West Dominick						
	Black River Boulevard and East Dominick St			2	1		1
	Did not cross a street	2					
total				6			5

Table 13
Pedestrian and bicycle counts: Intersection of North James Street and West Liberty Street
Source: Volpe Center

time:	intersection	Visitor Center / Fort ped	Visitor Center / Fort bike	Visitor Center / Fort total	Through site ped	Through site bike	Through site total
9:50-10:10	North James and West Dominick						
	Black River Boulevard and East Dominick St			0			2
	Did not cross a street				2		
10:50-11:10	North James and West Dominick	1			2		
	Black River Boulevard and East Dominick St			1			3
	Did not cross a street				1		
1:00-1:20	North James and West Dominick	2					
	Black River Boulevard and East Dominick St			2			0
	Did not cross a street						
total				3			5

Prioritized pedestrian recommendations

The following recommendations, in order of priority and shown on the map in Figure 47, are intended to improve pedestrian access and safety in and around the park.

1. *Develop partnerships and create a unified "Downtown Pedestrian Improvement Zone"*

   Fort Stanwix has an opportunity to connect each of Rome's three economic development corridors, which radiate from the park. By working with various stakeholders in each corridor, the park should consider spearheading an effort with the city to establish a pedestrian network for all of central Rome. As a single entity with multiple components, the Downtown Pedestrian Improvement Zone would find greatest success by securing outside assistance to fund pedestrian access and safety improvements throughout central Rome. This type of organization could also create incentives to provide connections to trails that travel outside the heart of the city.

2. *North James Street / West Dominick Street intersection improvements*

   Despite significant improvements in recent years, the intersection at North James Street and West Dominick Street remains unsafe for pedestrians. While the new textured and colored crosswalk is a step in the right direction, park staff and the study team find that high vehicle speeds through the intersection still pose a significant threat to pedestrians. Several alternatives for improving the safety of the intersection are listed below. Options include, in order of increasing expense:

   - Adding more signs to warn motorists of pedestrians in the area.
   - Creating a three-way stop at the intersection with two additional stop signs and road markings. Northbound and southbound traffic on North James Street will be required to come to a complete stop, in addition to traffic that is already required to stop when traveling from West Dominick Street onto North James Street.
   - Installing pedestrian-activated flashing lights that notify motorists when pedestrians are crossing the street.

- A raised crosswalk could be combined with any of the aforementioned alternatives. The crosswalk would be flush with existing sidewalks, serving as a physical high-speed deterrent as well as a psychological traffic-calming device. New signage to warn motorists of the raised pedestrian area would be required.
- Installing traffic lights complete with electronic pedestrian walk/do-not-walk signs.

The study team believes that the safest alternative will result from a complete stoppage of traffic, either with stop signs or a new traffic signal. The best choice can only be decided after a full traffic engineering analysis by the city's department of transportation, but a three-way stop sign is significantly less expensive to implement compared to some of the other alternatives above. Regardless of physical improvements, the city and park should work together to ensure that the state's pedestrian right-of-way law is adequately enforced at this intersection.

3. *Pedestrian-oriented wayfinding signage*

   As part of the development of a downtown pedestrian network, wayfinding signage at the intersections of major sidewalks and pathways can help visitors by providing distance and direction information to nearby sites. In addition to guiding people to Fort Stanwix from other parts of town, these signs could help direct park visitors to other attractions in the downtown area, including nearby business districts, the Fireman's Memorial on Black River Boulevard, the Tomb of the Unknown Soldier on North James Street and West Liberty Street, the Erie Canal, Bellamy Harbor Park, the Mohawk River, and visitor amenities in downtown Rome.

4. *Bicycle guidelines*

   The park may consider encouraging cyclists to dismount and walk the bicycle through the property. Currently, the paths are shared by pedestrians and bicyclists, and in some cases the size and speed of a bicycle is intimidating to walkers. Furthermore, the park has noticed an increase in erosion of earthworks on the site due to bicycles and pedestrians traveling off-trail. Signage would be necessary to implement this rule. Because non-motorized traffic flows vary significantly from season to season, guidelines should probably only be enforced during period of heavy visitation.

5. *Streetscape improvements along the park boundary*

   Improvements to the sidewalk along Black River Boulevard would benefit park visitors and community users. In particular, benches and lighting help to welcome visitors and allow city residents to enjoy the park not only as an historic national monument, but as public greenspace in the heart of Rome.

6. *Pathway improvements within the park boundary*

   Benches along the walkway between the visitor center and park entrance could offer visitors an opportunity to rest off the pathway. Benches should be of a similar design to the benches at the visitor center entrance and curve along the walkway and vegetated border.

7. *Other intersection improvements*

   Also in conjunction with a unified downtown pedestrian network, a number of additional crosswalks near the park should be completed at some of the busier intersections. New signalized crosswalks at Erie Boulevard and North James Street, Black River Boulevard and East Dominick Street, and Black River Boulevard and Court Street would improve pedestrian access and safety and encourage park visitors to walk into adjacent neighborhoods and business districts.

8. *Crosswalk maintenance*

   The park should work with the city to ensure existing pedestrian amenities are maintained. Currently, several crosswalks are significantly faded due to wear and tear.

Figure 47
Pedestrian safety recommendations
Source: Volpe Center

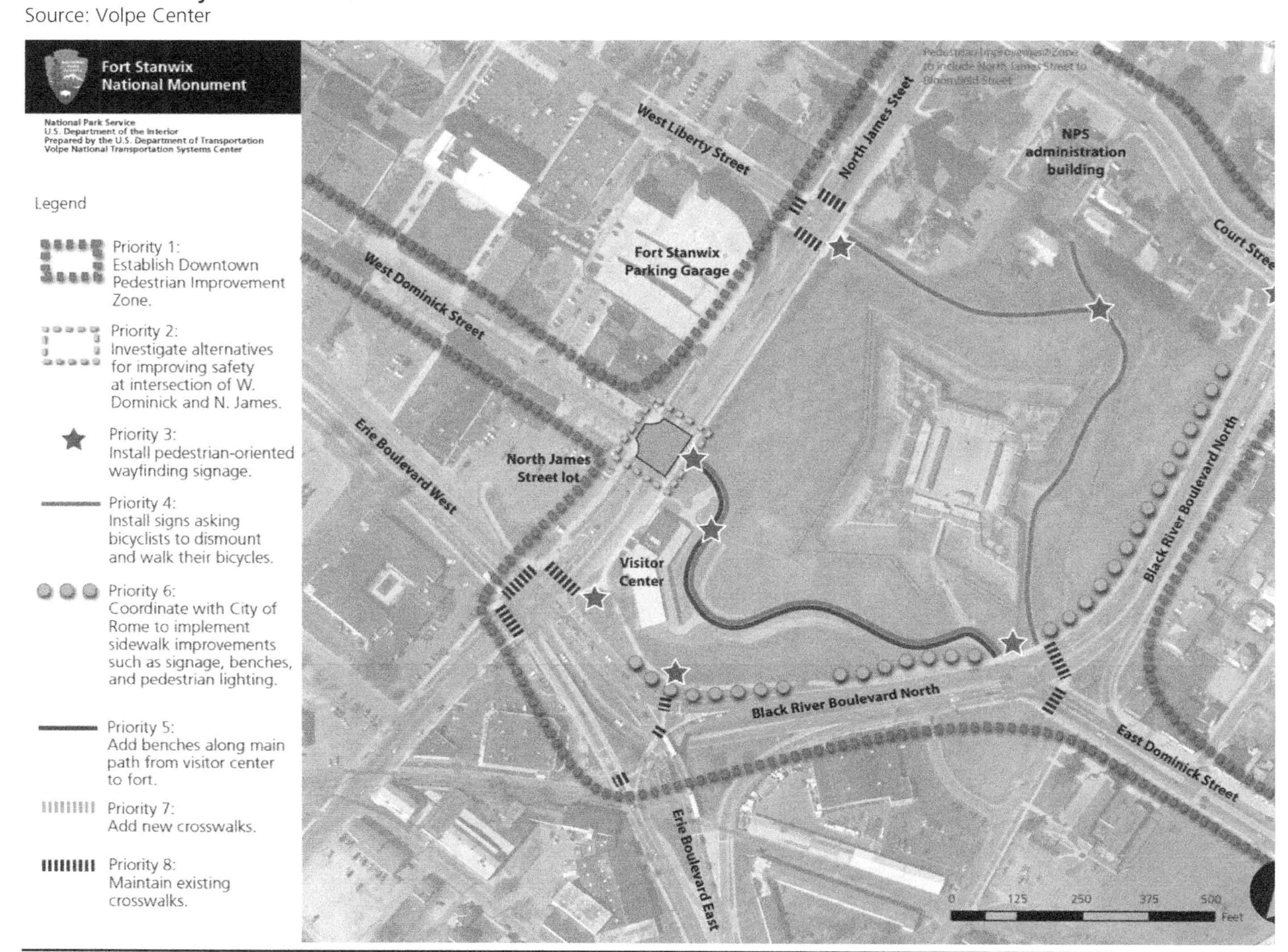

Shuttle opportunities

In April of 2008, NPS agreed to provide day-to-day management of Oriskany Battlefield State Historic Site. The battlefield is located just outside the town of Oriskany, NY, approximately six miles east of Fort Stanwix's location in downtown Rome, NY. The battle that took place at Oriskany is an important part of the Siege of Fort Stanwix, and NPS sees "a great opportunity to share resources and better connect the stories of these sites."[1]

As part of its cooperative agreement with the New York State Office of Parks, Recreation, and Historic Preservation, NPS rangers and maintenance employees work with state employees to provide site maintenance, interpretive programs, and other visitor services. To support the potential increase in programming at the battlefield and to better connect the battlefield to the park, the NPS is investigating potential alternative transportation options for travel between the two sites.

Purpose

The purpose of this section of the report is to (1) identify existing transportation options between Fort Stanwix and Oriskany Battlefield, (2) examine the potential benefits of a shuttle route, (3) explore the possibilities of land-based shuttles, water-based shuttles, and amphibious shuttles, (4) briefly discuss some of the major considerations for shuttle deployment, and (5) provide recommendations for shuttle alternatives.

Existing transportation options

Currently, the easiest way to travel from Fort Stanwix to Oriskany Battlefield is by car. The one-way distance is approximately six miles via NY- 69 East, a ten to fifteen minute drive. The only viable alternative transportation option is to travel by bicycle. While there is a possibility for an all-trail route between Fort Stanwix and Oriskany on the Canalway Trail in the future, existing biking conditions require significant portions of on-road travel. Potential improvements to the bike trail network are discussed in depth in the *Nonmotorized Trail Connections* section of this report.

Shuttle benefits

It is important to note that at time of this report's publishing, traffic congestion between Fort Stanwix and Oriskany is not heavy, and parking facilities at both sites are more-than-able to accommodate the visiting public on an average, summer weekend day. Consequently, the primary reasons to offer shuttle service between Fort Stanwix and Oriskany Battlefield are (1) to provide an alternative for those who prefer not to drive and (2) to increase visibility and visitation to Oriskany and better tie the battlefield to the park. Because the stories are so closely connected, it is conceivable that park visitors will be interested in experiencing the battlefield, particularly if transportation is provided.

Land-based shuttle

A land-based shuttle is the simplest option of motorized alternative transportation available to the park. A passenger van or low-capacity cutaway bus Figure 48) could be used, more than likely operating on only high visitation days, such as Saturdays and Sundays between Memorial Day and Labor Day.

---

[1] NPS Press Release. April 25, 2008. http://home.nps.gov/fost/parknews/fost-and-oriskany-battlefield-state-historic-site-sign-partnership-agreement.htm

Figure 48
Examples of cutaway buses
Source: Volpe Center

The primary advantage of the land-based shuttle alternative is that it is the least expensive way to provide transportation for larger groups to Oriskany Battlefield. Both capital and operating costs for a land-based shuttle are low (relative to other options discussed in this report), and the vehicle may serve other roles for the park during non-operational periods.

The disadvantage to a land-based shuttle is that it does not of itself contribute to the visitor experience, unless some interpretation is provided. The success of the shuttle would rest entirely on the draw of the destinations that it serves, and Oriskany Battlefield has not yet proven itself as a major attraction. Furthermore, from the visitor's perspective, a shuttle offers very little in the way of convenience when traffic and parking concerns are minimal at both sites.

A land-based shuttle route would likely follow a straightforward path to Oriskany Battlefield from the park, down Erie Boulevard and along NY 69 East (Figure 49).

Figure 49
Shuttle alternatives
Source: Volpe Center

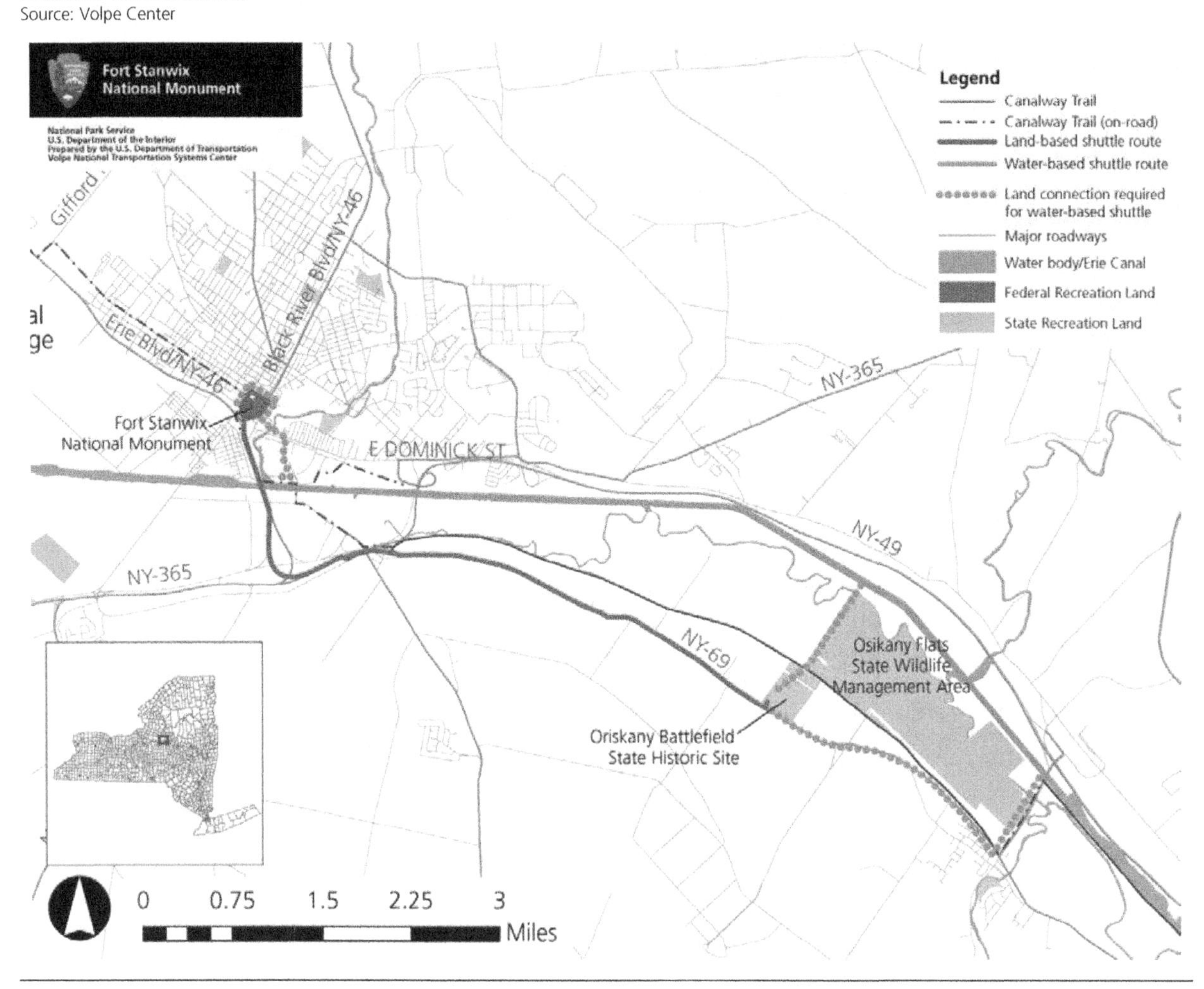

Water-based shuttle

Water-based transportation is a major part of Rome's history – during the colonial era as the location of the Great Portage and as a shipping center during the height of prominence of the Erie Canal. Today, recreational boaters are the primary users of the canal, and recent projections by the New York State Thruway Authority expect the number of recreational users to increase in the coming years. Based on discussions with the city of Rome and NPS officials, the Erie Canal is an integral component of the city's identity, both culturally and physically.

Since both Fort Stanwix and Oriskany Battlefield are each physically situated less than a mile (as the crow flies) from the Erie Canal, the study team examined the possibility of a water-based transportation system between the two sites. Fort Stanwix is about 0.8 miles from the closest passenger pick-up/drop-off point at Belamy Harbor Park. There are, however, no existing pick-up/drop-off facilities near Oriskany Battlefield (the nearest is approximately 5.3 on-road miles away at Lock 20 State Park), effectively eliminating the possibility of a water-based shuttle under existing circumstances.

It is worth mentioning that if a docking facility were located closer to the battlefield, the feasibility of a water-based shuttle would be greatly improved. Ground transportation between the respective pick-up/drop-off points and the actual destinations would still be a major consideration. Of note is the fact that currently the Oriskany Battlefield property abuts the state-owned Oriskany Flats

Wildlife Management Area. This publicly owned land could potentially provide a connection between the canal and the battlefield without any major property acquisition. Clearly, significant infrastructure improvements would be required – not the least of which includes a docking facility – but given the potential for development of an entirely new attraction as part of the Fort Stanwix / Oriskany Battlefield visitor experience, the study team has deemed it worthy of long-term consideration. An alternative passenger pick-up/drop-off docking facility may be considered at the River St/County Road 32 bridge, just north of the village of Oriskany. These options are illustrated in Figure 36.

Currently, several private organizations – both for- and non-profit – operate passenger boats on the Erie Canal (Figure 50). Most are tour boats that take passengers aboard the vessel for a prescribed time period (usually 1-2 hours) and provide narration to enhance the experience. Other passenger boat services include multi-day charters, boat rentals, and dinner cruises. The closest service to Fort Stanwix is based at the Erie Canal Village, just west of Rome. The vessel does not actually operate on the main channel of the existing barge canal; rather, it operates on an unused portion of the old Erie Canal and is powered entirely by a team of horses.

Figure 50
Erie Canal tour boats
Sources: http://eriecanalcruises.com/clips.html; http://www.samandmary.org/index.php?cat=cruises&page=sampatch

Amphibious shuttle

The final shuttle consideration examines the possibility of using an amphibious vehicle to provide water-based shuttle service between Fort Stanwix and Oriskany Battlefield, thus eliminating the need for separate land-based vehicles connecting the canal with the destinations. While the concept is less conventional than a traditional land- or water-based shuttle, amphibious vehicles are currently being used in cities throughout the U.S. by private-sector tour operators. Often called "duck tours," the closest operating amphibious tour service is located in Albany, while other companies operate in eastern cities such as Boston, Philadelphia, Baltimore, Washington, and Pittsburgh. Also, a National Park-oriented amphibious tour service is offered in Hot Springs, Arkansas by an independent operator (Figure 51).

In order to launch, an amphibious vehicle requires a concrete boat ramp. A boat ramp is available close to Fort Stanwix at the corner of South James Street and Muck Road, but again, there are no facilities within a reasonable distance of Oriskany Battlefield. Again, a facility would have to be constructed, and the site with the most potential is likely the River Street / County Road 32 bridge near Oriskany Village.

Figure 51
Amphibious vessel in Hot Springs, Arkansas
Source: http://www.rideaduck.com/ducks/index.php?p=home

General shuttle considerations

If the park wishes to pursue shuttle service – regardless of whether it is land-based, water-based, or both – numerous cost-determining factors such as service characteristics, vehicle design, and the overall model of operations, must be considered.

Service characteristics

A detailed service plan is essential for understanding the day-to-day cost of operating the shuttle. Variables include dates and days of operation, shuttle headway (time between vehicles), route locations, stop locations, and dwell time, all of which can be adjusted in order to minimize costs while optimizing the services offered.

Vehicle type and design

The type of service and projected ridership will help determine the necessary vehicle characteristics. Depending on estimated number of passengers on each trip and whether the vehicle will be traveling on land, water, or both, vehicle design will play an important role in acquisition costs, fuel costs, licensing requirements, maintenance, and visitor comfort and experience.

Operations Model

A key decision concerns whether NPS will operate the shuttle service in-house or work with a third-party contractor. There are advantages and disadvantages to both alternatives. Operating independently may be less expensive than a private contract and may allow the park to take advantage of special opportunities like specific grant programs or benefit from partnerships with other organizations. On the other hand, a private provider with more knowledge and experience in providing shuttle services may be able to do so more efficiently by using existing vehicle and staff resources. A private contract may also provide the park with more flexibility. The contract can be organized such that the contractor is responsible for any combination of system operations and performance requirements, including staffing, monitoring, vehicle procurement, vehicle maintenance, and vehicle storage.

Shuttle recommendations

At this point in time, the study team does not recommend that Fort Stanwix pursue a shuttle service to Oriskany Battlefield. In order to reconsider this recommendation, visitation data for Oriskany Battlefield will have to be accurately monitored in order to provide a better understanding of ridership potential. The demand for such a shuttle service does not appear to be strong enough.

The study team does support any short-term land-based pilot opportunities, for which a test shuttle service is implemented for one season with borrowed vehicles and temporary staffing. This type of project would allow NPS to assess the viability of a Fort Stanwix – Oriskany shuttle with minimal capital investment. Based on the project's success, NPS and the park will have a better understanding of the pattern of visitation to both Fort Stanwix and Oriskany Battlefield and resulting shuttle service needs.

While a water-based or amphibious shuttle is far less feasible at this point in time, the study team believes that, if logistically feasible, this type of service may experience higher ridership potential than a land-based shuttle. The reasoning is that a water-based shuttle service may be seen as an attraction in and of itself, entertaining and educating riders while en route to the destinations served.

Unfortunately, pilot opportunities for a water-based or amphibious shuttle will be difficult to pursue due to the lack of a docking facility and/or passenger pick-up/drop-off point near Oriskany Battlefield. Should conditions change in a way that favors this type of service, the study team recommends a preliminary pilot project similar to the one mentioned above for land-based shuttles. Existing canal tour operators may be willing to assist in this process.

REPORT DOCUMENTATION PAGE	*Form Approved* *OMB No. 0704-0188*

1. REPORT DATE *(DD-MM-YYYY)*	2. REPORT TYPE	3. DATES COVERED *(From - To)*
xx-11-2011	Final	May 2009 - Nov 2011

4. TITLE AND SUBTITLE
Fort Stanwix National Monument:
Alternative Transportation Study

5a. CONTRACT NUMBER F4505087777
5b. GRANT NUMBER
5c. PROGRAM ELEMENT NUMBER

6. AUTHOR(S)
Ben Rasmussen; Ben Cotton; Kirsten Holder

5d. PROJECT NUMBER NP69
5e. TASK NUMBER KLG31
5f. WORK UNIT NUMBER

7. PERFORMING ORGANIZATION NAME(S) AND ADDRESS(ES)
U.S. Department of Transportation
Research and Innovative Technology Administration
John A. Volpe National Transportation Systems Center
55 Broadway, Cambridge, MA 02142

8. PERFORMING ORGANIZATION REPORT NUMBER
DOT-VNTSC-NPS-12-07

9. SPONSORING/MONITORING AGENCY NAME(S) AND ADDRESS(ES)
NPS Northeast Region
15 State St.
Boston, MA 02109

10. SPONSOR/MONITOR'S ACRONYM(S)
NPS NER

11. SPONSOR/MONITOR'S REPORT NUMBER(S)
015/115586

12. DISTRIBUTION/AVAILABILITY STATEMENT
Public distribution/availability

13. SUPPLEMENTARY NOTES

14. ABSTRACT
As a follow-up to Volpe's 2006 Transportation Summary Report, this project addresses a number of specific transportation concerns in Fort Stanwix National Monument. In addition to an update of the 2006 existing conditions report to reflect recent changes that have occurred in and around Fort Stanwix, this report focuses on five areas of primary interest to the fort: (1) nonmotorized trail connections, (2) vehicular signage and wayfinding, (3) parking, (4) pedestrian access, and (5) shuttle feasibility.

15. SUBJECT TERMS
Fort Stanwix; alternative transportation; NPS; signage and wayfinding; parking; pedestrian access and safety; shuttle

16. SECURITY CLASSIFICATION OF:			17. LIMITATION OF ABSTRACT	18. NUMBER OF PAGES	19a. NAME OF RESPONSIBLE PERSON
a. REPORT	b. ABSTRACT	c. THIS PAGE			Ben Rasmussen
None	None	None	N/A		19b. TELEPHONE NUMBER *(Include area code)* 617-494-2768

Reset **Standard Form 298** (Rev. 8/98)
Prescribed by ANSI Std. Z39.18

Made in the USA
Coppell, TX
05 October 2021